Lecture Notes in Mathematics

Edited by A. Dold and B. Eckmann

436

Louis Auslander
Richard Tolimieri

Abelian Harmonic Analysis, Theta Functions and Function Algebras on a Nilmanifold

Springer-Verlag
Berlin · Heidelberg · New York 1975

Prof. Louis Auslander
The Graduate School and University Center
City University of New York
33 West, 42 Street
New York, N.Y. 10036/USA

Prof. Richard Tolimieri
Dept. of Mathematics
University of Connecticut
Storrs, CT 06250/USA

Library of Congress Cataloging in Publication Data

Auslander, Louis.
Abelian harmonic analysis, theta functions, and functional analysis on a nilmanifold.

(Lecture notes in mathematics ; 436)
Bibliography: p.
Includes index.
1. Lie groups. 2. Manifolds (Mathematics) 3. Harmonic analysis. 4. Functions, Theta. I. Tolimieri, Richard, 1940- joint author. II. Title. III. Series: Lecture notes in mathematics (Berlin) ; 436.
QA3.L28 no.436 [QA387] 510'.8s [512'.55]
74-32366

AMS Subject Classifications (1970): 22EXX, 22E25, 43-XX, 43A85

ISBN 3-540-07134-2 Springer-Verlag Berlin · Heidelberg · New York
ISBN 0-387-07134-2 Springer-Verlag New York · Heidelberg · Berlin

Offsetdruck: Julius Beltz, Hemsbach/Bergstr.

5 ep

PREFACE

These notes are concerned with the inter-relationship between abelian harmonic analysis, theta functions and functional analysis on a certain nilmanifold. Some of the results in these notes are new and some are old. However, our approach, because it puts a certain nilmanifold and its function theory at center stage, often leads to new proofs of standard results. For example, we view theta functions as the analogue on a nilmanifold of the spherical functions on the sphere, where the Heisenberg group plays the role of the orthogonal group. Thus the classical theta identities will follow from basic group theoretic results.

Historically, there are many names that can be associated with the topics treated in these notes. Because of the informal nature of these notes we have not made any effort at giving complete biographical references for results and have given only references to the sources we ourselves have used. If we have overlooked someone's work or state a result without reference that someone knows to be his, we apologize in advance. However, we would be less than honest if we did not admit the great influence of the ideas of J.Brezin, G.W.Mackey and A.Weil on our work. Indeed, after so many years of talking with Brezin many of the ideas or germs of ideas in these notes may be his . In addition, we should also mention the work of Cartier which stands somewhere in the middle ground between the work of Weil and that presented in these notes.

It may be advisable at this point to explain to the reader how the material in these notes has been labelled. The reader will find some material labelled Chapter and some material labelled Appendix. The material with sections labelled Chapter is more complete and well rounded. The material labelled Appendix is of a more tentative nature and does not yet seem to have taken on a definitive form. The last appendix has its own list of references and the other appendices and chapters have a single list of references located on page 74.

TABLE OF CONTENTS

CHAPTER I

THE FOURIER TRANSFORM AND THE NILMANIFOLD $\Gamma\backslash N$.

The deep relationship between abelian harmonic analysis, Jacobi theta functions, and the function theory on a certain nilmanifold that is the subject of these notes can be well motivated by examining a non-traditional proof of the Plancherel theorem for the reals. Hints of the proof we will present can be found in the proof of the general Plancherel theorem given by A. Weil in [8]. We will begin by giving a formal treatment that ignores all convergence questions.

Let $\mathbb{R}$ denote the reals and let $f(t)$ be a function on $\mathbb{R}$. We define the functions $f_n(\xi)$, $0 \leq \xi < 1$, $n \in \mathbf{Z}$, where $\mathbf{Z}$ denotes the integers, by the formula

$$f_n(\xi) = f(n + \xi) \qquad 0 \leq \xi < 1, \quad n \in \mathbf{Z}.$$

Now let $\mathbf{T}$ denote the reals modulo 1 and let $\mathbf{T}^2 = \mathbf{T} \times \mathbf{T}$ denote the 2-torus. We, of course, may view $f_n(\xi)$ as a function on $\mathbf{T}$ for each value of $n \in \mathbf{Z}$. Define Θf by the formula

$$\Theta f = \sum_{m=-\infty}^{\infty} f_m(\xi) \exp(2\pi i m\eta) , \quad 0 \leq \eta < 1$$

and view Θf as a formal sum on $\mathbf{T}^2$. Here (ξ,η) is being considered as a coordinate system for $\mathbf{T}^2$. Let

$$U_A \Theta f = \sum_{m=-\infty}^{\infty} f_m(\eta) \exp(-2\pi i m\xi) .$$

Since $0 \leq \eta < 1$ and $0 \leq \xi < 1$, the function $\exp(-2\pi i\eta\xi)$ is well defined, although discontinuous, on $\mathbf{T}^2$. Thus we may define

$$M\, U_A\, \Theta f = \sum_{m=-\infty}^{\infty} \exp(-2\pi i\eta\xi) f_m(\eta) \exp(-2\pi i m\xi)$$

and finally define

$$\hat{f}(-2\pi(\xi+n)) = \int_0^1 M\, U_A\, \Theta(f) \exp(-2\pi i n\eta)\, d\eta$$

$n \in \mathbf{Z}$. Letting $t = \xi + n$, we have $\hat{f}(-2\pi t) = \hat{f}(-2\pi(\xi+n))$.

We will now formally compute $\hat{f}(-2\pi t)$.

$$\hat{f}(-2\pi(\xi+n)) = \int_0^1 \sum_{m=-\infty}^{\infty} \exp(-2\pi i\eta\xi) f_m(\eta) \exp(-2\pi i m\xi) \exp(-2\pi i n\eta)\, d\eta$$

$$= \sum_{m=-\infty}^{\infty} \int_0^1 \exp(-2\pi i\eta\xi) f_m(\eta) \exp(-2\pi i m\xi) \exp(-2\pi i n\eta)\, d\eta\ .$$

Setting $s = \eta + m$, we have, because n and m are in $\mathbf{Z}$, that

$$\hat{f}(-2\pi t) = \int_{-\infty}^{\infty} f(s) \exp(-2\pi i st)\, ds\ .$$

Let us now see what all this formalism means when $f \in L^2(\mathbb{R})$.

We begin our discussion by showing that Θ is an isomorphism of $L^2(\mathbb{R})$ onto $L^2(\mathbf{T}^2)$. First, note that for $f \in L^2(\mathbb{R})$

$$\sum_{m=-\infty}^{\infty} \|f_m\|^2_{L^2(\mathbf{T})} = \|f\|^2_{L^2(\mathbb{R})}$$

where $\|\ \|$ denotes the norm and the subscript the Hilbert space in which the function lies. Next observe that

$$f_m(\xi) \exp 2\pi i m\eta \quad \text{and} \quad f_n(\xi) \exp 2\pi i n\eta\ , \quad n \neq m$$

are orthogonal in $L^2(\mathbf{T}^2)$. Hence

$$\|\Theta f\|^2_{L^2(\mathbf{T}^2)} = \sum_{m=-\infty}^{\infty} \|f_m\|^2_{L^2(\mathbf{T}^2)} = \sum_{m=-\infty}^{\infty} \|f_m\|^2_{L^2(\mathbf{T})} = \|f\|^2_{L^2(\mathbb{R})}\ .$$

This shows that $\Theta\colon L^2(\mathbb{R}) \to L^2(\mathbf{T}^2)$ is an isometry.

We will now verify that Θ is a surjection. Let $F(\xi,\eta) \in L^2(\mathbf{T}^2)$ and let

$$\sum a_{nm} \exp(2\pi i n\eta) \exp(2\pi i m\xi)$$

be the Fourier series for $F(\xi,\eta)$. Then

$$\|F\|^2_{L^2(\mathbf{T}^2)} = \sum_{n,m=-\infty}^{\infty} |a_{nm}|^2$$

and $\Sigma_{n=-N}^{N} a_{nm} \exp 2\pi i m\xi$ converges in L^2 to $f_n(\xi)$ with

$$\|f_n(\xi)\|^2_{L^2(\mathbf{T})} = \sum_{m=-\infty}^{\infty} |a_{nm}|^2\ .$$

Now let $f(x) = f_n(\xi)$, where $x = n + \xi$, be considered as a function on $\mathbb{R}$. We note that if $\psi F = f$ then

$$\|\psi F\|_{L^2(\mathbf{T}^2)} = \|f\|_{L^2(\mathbb{R})}$$

because

$$\|f\|^2_{L^2(\mathbb{R})} = \sum_{n=-\infty}^{\infty} \|f_n(\xi)\|^2_{L^2(\mathbf{T})} = \sum_{n,m} |a_{nm}|^2 = \|F\|^2_{L^2(\mathbf{T}^2)} .$$

Thus $\psi: L^2(\mathbf{T}^2) \to L^2(\mathbb{R})$ is an isometry. It is a formal computation to show that $\psi = \Theta^{-1}$ and so Θ is a surjection. We have thus shown that Θ is a unitary operator.

Now let $F \in L^2(\mathbf{T}^2)$. Then we may view F as a doubly periodic function on $\mathbb{R}^2$ or $F(x,y)$ is such that

$$F(x+n, y+m) = F(x,y) , \quad n,m \in Z \quad (x,y) \in \mathbb{R}^2 .$$

We define $U_A F(x,y) = F(y,-x)$. Clearly

$$\|U_A F\|_{L^2(\mathbf{T}^2)} = \|F\|_{L^2(\mathbf{T}^2)}$$

and as U_A is invertable U_A is a unitary operator.

Now let $F \in L^2(\mathbf{T}^2)$, $1 \le p$, and let

$$(MF)(\xi,\eta) = \exp(-2\pi i \xi\eta) F(\xi,\eta) , \quad 0 \le \xi,\eta < 1 .$$

So again

$$\|MF\|_{L^2(\mathbf{T}^2)} = \|F\|_{L^2(\mathbf{T}^2)}$$

and so M being invertable is a unitary operator of $L^2(\mathbf{T}^2)$.

This proves that

$$\Theta^{-1} M U_A \Theta : L^2(\mathbb{R}) \to L^2(\mathbb{R})$$

is a unitary operator. We will let $\mathcal{J} = \Theta^{-1} M U_A \Theta$.

Let the Fourier expansion of $M U_A \Theta f = \sum_{m,n=-\infty}^{\infty} b_{mn} \exp 2\pi i m\xi \exp 2\pi i n\eta$ where convergence is in the L^2 norm. Then $\hat{f}(-2\pi(\xi+n)) = \sum_{m=-\infty}^{\infty} b_{mn} \exp 2\pi i m\xi$, where convergence is in the L^2 norm on $\mathbf{T}$. We next note that if g_n and

$g \in L^2(T)$, $n = 1,2,\ldots$, and if A_{mn} and A_m is the m Fourier coefficient of g_n and g, respectively, then if $g_n \to g$ in $L^2(T)$ we have $\lim_{n\to\infty} A_{mn} = A_m$. This shows that for $f \in L^2(\mathbb{R})$, we have

$$\hat{f}(-2\pi t) = \lim_{N\to\infty} \int_{-N}^{N} f(s)\exp(-2\pi ist)ds$$

where the limit is in the L^2 norm. Thus

$$\mathcal{J} : L^2(\mathbb{R}) \to L^2(\mathbb{R})$$

is a unitary operator. This last statement is called Plancherel's theorem. Since $U_A F(0,0) = F(0,0)$ and $\exp(-2\pi i\xi\eta)(0,0) = 1$, we have

$$\sum_{m=-\infty}^{\infty} f(m) = \sum_{m=-\infty}^{\infty} \hat{f}(2\pi m)$$

when both $M\,U_A\,\Theta f$ and $\Theta\hat{f}$ are continuous at $(0,0)$. This is called the Poisson summation formula.

The proof of the Plancherel theorem we have just presented is elementary, but seems very artificial. Let us now present a more geometric treatment of the above proof that will lead us to an explaination of why it works.

Consider the group N, often called the 3-dimensional Heisenberg group, of all matrices of the form

$$\begin{pmatrix} 1 & y & z \\ 0 & 1 & x \\ 0 & 0 & 1 \end{pmatrix}$$

where $x,y,z \in \mathbb{R}$. We will denote the elements of N by triples (x,y,z) and then multiplication is given by

$$(x,y,z)(a,b,c) = (x+a,y+b,z+c+ya) \quad .$$

In N let Γ denote the discrete subgroup consisting of all triples (m_1,m_2,m_3), $m_i \in Z$, $i = 1,2,3$. Then the homogeneous space $\Gamma\backslash N$ consisting of cosets of the form Γn, $n \in N$, is a compact manifold, called a nilmanifold. Further N has a left and right invariant Haar measure. Hence there exists a unique probability measure μ on $\Gamma\backslash N$ which is invariant under the action of N on $\Gamma\backslash N$ given by $\Gamma n \to \Gamma n n_0$, $n, n_0 \in N$. We next note that the set

$$D = \{(x,y,z) \in N \mid 0 \le x,y,z < 1\}$$

is a fundamental domain for $\Gamma\backslash N$ and that μ is given in this coordinate system by the usual Lebesgue measure.

Let us now consider $L^2(\Gamma\backslash N)$ and for $f \in L^2(\Gamma\backslash N)$ and $n_1, n_2 \in N$, we define

$$R(n_1)(f(\Gamma n_2)) = f(\Gamma n_2 n_1) .$$

Then the mapping $R: n \to R(n)$ is a unitary representation of N with kernel $C = \{(0,0,m) \in N \mid m \in Z\}$. This gives us a representation of N/C. The image of $(0,0,z) \in N$, $z \in \mathbf{R}$ in N/C is then a central compact subgroup and so $L^2(\Gamma\backslash N)$ breaks up into a Hilbert space direct sum

$$L^2(\Gamma\backslash N) = \oplus \sum_{n \in Z} H_n$$

where $F \in H_n$ if and only if

$$R(0,0,z)F = e^{2\pi i n z} F .$$

If p_n is the orthogonal projection of $L^2(\Gamma\backslash N)$ onto H_n, then we may write p_n explicitly in the form

$$p_n(F)(x,y,z) = \int_0^1 F(x,y,z+t)e^{-2\pi i n t}\, dt , \quad n \in Z .$$

It is clear that the spaces H_n, $n \in Z$, are invariant under the representation R of N.

We will now look a bit harder at H_n and produce an orthonormal basis for H_n. Because D is a fundamental domain for $\Gamma\backslash N$ and the measure defining $L^2(\Gamma\backslash N)$ agrees with the Lebesgue measure on D, one easily verifies that

$$\exp(2\pi i \alpha_1 x)\exp(2\pi i \alpha_2 y)\exp(2\pi i \alpha_3 z) , \quad \alpha_1, \alpha_2, \alpha_3 \in Z$$

$0 \le x,y,z < 1$ form an orthonormal basis of $L^2(\Gamma\backslash N)$. In terms of this basis of $L^2(\Gamma\backslash N)$, H_n has orthonormal basis all basis vectors with $\alpha_3 = n$. Thus the mapping

$$\mathcal{J}_n : L^2(\mathbf{T}^2) \to H_n$$

defined on an orthonormal basis of $L^2(\mathbf{T}^2)$ by

$$\mathcal{J}_n(\exp(2\pi i\alpha_1 x)\ \exp(2\pi i\alpha_2 y)) = \exp(2\pi i\alpha_1 x)(\exp\ 2\pi i\alpha_2 y)(\exp(2\pi inz)$$

extends to a unitary operator.

The space H_1 will play the role of $L^2(\mathbf{T}^2)$ in our geometric interpretation of our proof of the Plancherel theorem. It remains to find a geometric meaning for the operators MU_A and Θ. We will begin with the operator MU_A.

Let $L(N)$ denote the Lie algebra of N. Then $L(N)$ consists of all matric of the form

$$\begin{pmatrix} 0 & b & c \\ 0 & 0 & a \\ 0 & 0 & 0 \end{pmatrix} \qquad a,b,c \in \mathbb{R}$$

with the usual bracket relation for making these matrices into a Lie algebra. Le

$$X = \begin{pmatrix} 0 & 0 & 0 \\ 0 & 0 & 1 \\ 0 & 0 & 0 \end{pmatrix} \quad Y = \begin{pmatrix} 0 & 1 & 0 \\ 0 & 0 & 0 \\ 0 & 0 & 0 \end{pmatrix} \quad Z = \begin{pmatrix} 0 & 0 & 1 \\ 0 & 0 & 0 \\ 0 & 0 & 0 \end{pmatrix}$$

Then X,Y,Z is a basis of $L(N)$ and Z spans the center of $L(N)$. Further

$$[Y,X] = YX - XY = Z \quad .$$

Now let $\mathfrak{a}^*$ be the linear mapping of $L(N)$ given by

$$\mathfrak{a}^*(Y) = X\ ,\ \mathfrak{a}^*(X) = -Y \text{ and } \mathfrak{a}^*(Z) = Z\ .$$

Then

$$[\mathfrak{a}^*(Y),\mathfrak{a}^*(X)] = [\ X,-Y] = [Y,X] = Z$$

and so $\mathfrak{a}^*$ is an automorphism of $L(N)$ and so, since N is simply connected, $\mathfrak{a}^*$ determines an automorphism $\mathfrak{a}$ of N. We will by direct computation now compute $\mathfrak{a}$. First note that

$$(x,y,z) = (x,0,0)(0,y,0)(0,0,z) \quad .$$

If $\exp(\ )$ denotes the exponential map of $L(N)$ into N then $\exp(xX) = (x,0,($ etc. Hence

$$\alpha(x,y,z) = \alpha(x,0,0)\ \alpha(0,y,0)\ \alpha(0,0,z)$$
$$= (0,-x,0)(y,0,0)(0,0,z)$$
$$= (y,-x,-xy+z)\quad .$$

Thus α is an automorphism of N that maps Γ onto itself and so α induces a 1-1 measure preserving mapping

$$\alpha^{\#} : \Gamma\backslash N \to \Gamma\backslash N\quad .$$

Hence $\alpha^{\#}$ determines a unitary operator of $L^2(\Gamma\backslash N)$ which, because α acts trivially on the center of N, maps H_n onto itself and so maps H_1 onto itself. Let this unitary operator on H_1 be denoted by $\alpha_U^{\#}$. We will now explicitly compute the operator $\alpha_U^{\#}$. Let

$$F(x,y,z) = \exp(2\pi i\alpha_1 x)\ \exp(2\pi i\alpha_2 y)\ \exp(2\pi iz)$$

then

$$(\alpha_U^{\#}F)(x,y,z) = \exp(2\pi i\alpha_1 y)\ \exp(-2\pi i\alpha_2 x)\ \exp(2\pi i(z-xy))\quad .$$

This proves that

$$\mathcal{J}^{-1}\alpha_U^{\#}\mathcal{J} = MU_A$$

and gives a geometric meaning to the operator MU_A. Thus we have proven that

$$\hat{f}(-2\pi t) = \Theta^{-1}\mathcal{J}^{-1}\alpha_U^{\#}\mathcal{J}\Theta f(t)\ ,\qquad f \in L^2(\mathbb{R})\ .$$

There still remains the task of understanding the unitary operator $\mathcal{J}\Theta : L^2(\mathbb{R}) \to H_1$. We will first give a characterization of $\mathcal{J}\Theta$ and then look deeper as to why $\mathcal{J}\Theta$ works. In order to do this we begin by giving a unitary representation of N on $L^2(\mathbb{R})$. To do this, let $f(t) \in L^2(\mathbb{R})$ and define for $x,y,z \in \mathbb{R}$ the unitary operator

$$U(x,y,z)\ f(t) = e^{2\pi i(z+tx)}\ f(t+y)\quad .$$

If we now compute $U(x,y,z)\ U(a,b,c)$ an elementary computation gives that this is $U(x+a,y+b,z+c+ya)$ and so

$$U : (x,y,z) \to U(x,y,z)$$

defines a unitary representation of N. By classical results this representation is irreducible.

Now let us view

$$F(x,y,z) = \sum_n f(n+y)\, e^{2\pi inz}\, e^{2\pi iz}$$

as a function on N. Then $F((n_1,n_2,n_3)(x,y,z)) = F(x,y,z)$ because

$$(n_1,n_2,n_3)(x,y,z) = (n_1+x,n_2+y,z+n_3+n_2x)$$

and so

$$F((n_1,n_2,n_3)(x,y,z)) = \sum_n f(n+n_2+y)\, e^{2\pi in(x+n_1)}\, e^{2\pi i(z+n_2x)}$$

$$= \sum_n f(n+n_2+y)\, e^{2\pi i(n+n_2)x}\, e^{2\pi iz}$$

and so letting $m = n + n_2$ we have the desired result.

We next want to compute $(R(a,b,c)F)(x,y,z)$. To do this note that $(x,y,z)(a,b,c) = (x+a,y+b,z+c+ya)$ and so

$$(R(a,b,c)F)(x,y,z) = e^{2\pi ic}\, e^{2\pi iz} \sum f(n+b+y)\, e^{2\pi in(x+a)}\, e^{2\pi iya} .$$

Now

$$\mathcal{J}\,\Theta\, U(a,b,c)f = \mathcal{J}\,\Theta(e^{2\pi i(c+ta)} f(t+b))$$

$$= e^{2\pi iz} \sum_n e^{2\pi inx}\, e^{2\pi i(c+(n+y)a)} f(n+b+y) .$$

Thus

$$\mathcal{J}\,\Theta\, U = R\,\mathcal{J}\,\Theta$$

and we have shown that $\mathcal{J}\Theta$ is an intertwining operator between R and an irreducible unitary representation U of N and hence R is an irreducible representation of N. Thus $\mathcal{J}\Theta$ being an intertwining operator between irreducible representations is essentially unique by Shur's lemma. We have thus characterized the unitary operator $\mathcal{J}\Theta$, but unfortunately, this characterization sheds little light on why $\mathcal{J}\Theta$ works as the desired intertwining operator.

In order to understand why $\mathcal{J}\Theta$ works, we must make a detour through enough of the Mackey machine to show us the geometric pictures behind the mapping $\mathcal{J}\Theta$. This detour will also have the advantage of providing us with an out-line of a proof of the fact that H_n is a multiplicity space for the representation R of multiplicity $|n|$, a result that will play an important role in our later work. Since J. Brezin is preparing a detailed monograph on the general results in this direction, we will content ourselves with showing how special cases of the general results can be used to carry out the computations we need. We will state our special results as theorems, but not present proofs.

In N consider the abelian normal subgroup

$$B = \{(x,0,z) \in N \mid x,z \in \mathbb{R}\}$$

and let $\Gamma \cap B = L$. Let $B^\wedge$ be the dual group of B. Let Y be the subgroup of N consisting of all elements of the form $(0,y,0)$, $y \in \mathbb{R}$. Then N is the semi-direct product of Y and B. Since B is abelian N acting on B by inner automorphisms is equivalent to Y acting on B by inner automorphisms. Now

$$(0,y,0)(x,0,z)(0,-y,0) = (x,0,z+yx) .$$

Thus if we view Y acting on $B^\wedge$, we have

$$y(e^{2\pi i(\alpha x+\beta z)}) = \exp^{2\pi i[(\alpha+\beta y)x+\beta z]} , \quad \alpha,\beta \in \mathbb{R} .$$

Hence if $\beta \neq 0$, Y acts with trivial isotropy group.

Now the regular representation of B on $L^2(L\backslash B)$ breaks up into a Hilbert space direct sum

$$L^2(L\backslash B) = \oplus \sum_{m_1,m_3 \in \mathbb{Z}} \mathbb{C}\, e^{2\pi i(m_1 x+m_3 z)}$$

where $\mathbb{C}$ denotes the complex numbers.

Now let $N^\wedge$ denote the Mackey dual of the group N. By definition, $N^\wedge$ is the set of all unitary equivalence classes of irreducible unitary representation of N endowed with a certain Borel structure. Since we will have no need to use this Borel structure in the rest of these notes, we will not bother to define it. Since $\Gamma\backslash N$ is compact it follows from general results that there exist, R-invariant subspaces $\theta(r) \in L^2(\Gamma\backslash N)$ for $r \in N^\wedge$ such that $R|\theta(r)$ is of finite multiplicity and

$$L^2(\Gamma\backslash N) = \oplus \sum_{r \in (\Gamma\backslash N)^\wedge} \theta(r)$$

where $(\Gamma\backslash N)^\wedge$ is some countable subset of $N^\wedge$.

In our case the space H_0 and H_n ; $n \neq 0$, must be treated separately. But the case H_0 consists of functions on $\Gamma\backslash N$ that are constant along the cosets of the center and H_0 may be identified with the decomposition of a two dimensional torus. Or, more precisely,

$$H_0 = \oplus \sum_{m_1, m_2 \in \mathbb{Z}} \mathbb{C} \exp 2\pi i(m_1 x + m_2 y) \quad .$$

Hence the interesting new case is when $n \neq 0$ which we will now treat in some detail.

Let $B_1^\wedge$ be the characters on B with $m_3 \neq 0$ and let $(\Gamma\backslash N)_1^\wedge$ be the subset of $(\Gamma\backslash N)^\wedge$ which are not of the form $\exp 2\pi i(m_1 x + m_2 y)$. It requires a modification of an argument of Mackey, but one can prove

<u>Theorem 1</u>: For each $r \in (\Gamma\backslash N)_1^\wedge$ there exists $q \in B_1^\wedge$, i.e., a character of the form $\exp 2\pi i(m_1 x + m_3 z)$, $m_3 \neq 0$, such that inducing q to N yields r.

Thus to describe the space $\theta(r)$ in terms of the characters $\exp 2\pi i(m_1 x + m_3 z)$ we have to be able to describe the representation induced from the character which we will denote by $I_B^N(m_1, m_3)$, $m_3 \neq 0$. The usual space on which to realize $I_B^N(m_1, m_3)$ is

$$L^2(N/B) = L^2(\mathbb{R}) \quad .$$

Now we may identify the range of a function in $L^2(N/B)$ with $\mathbb{C} \exp 2\pi i(m_1 x + m_3 z) \subset L^2(B/L)$. Hence we may view

$$L^2(N/B) \subseteq L^2(N/B, L^2(B/L))$$

where the right hand side denotes mappings from N/B to $L^2(B/L)$. We now get an isometric isomorphism

$$T : L^2(N/B, L^2(B/L)) \to L^2(N/L)$$

simply by noting that an element of $L^2(N/B, L^2(B/L))$ assigns to each $y \in Y$ a function on B/L. Thus, since

$$L\backslash N = Y \times B/L ,$$

to each point in $(Y \times B/L)$ it assigns a number and hence is a function on $L\backslash N$. The statements about norms are then easily verified. But, we are not yet done, for we have only replaced $L^2(N/B)$ by $L^2(L\backslash N)$, and what we want is $L^2(\Gamma\backslash N)$. We now define

$$A : L^2(L\backslash N) \to L^2(\Gamma\backslash N)$$

by

$$(AF)(\Gamma g) = \Sigma_{x\in\Gamma/L} F(xg)$$

where F is a bounded function on $L\backslash N$ with compact support.

Theorem 2: The composition AT defines an isometry from $L^2(N/B)$ into a possibly proper subspace $\varphi^{\#}(p)$ of $\theta(r)$. Further, $\varphi^{\#}(p)$ is R-invariant.

We will verify in a little while that $AT = \mathcal{J}\Theta$ when p has the form $\exp 2\pi i(n_1 x+z)$.

There are apt to be many elements of $(B/L)^{\wedge}$ that induce r, and that is why $\varphi^{\#}(p)$ need not be all of $\theta(r)$. Let

$$S(r) = \{p \in (B/L)^{\wedge} | I_B^N(p) = r\} .$$

It can be verified that if $\gamma \in \Gamma$ and $p \in S(r)$ then $\gamma(p) \in S(r)$. Hence Γ acts on $S(r)$.

Theorem 3: Let p and $q \in S(r)$. If $p \in \Gamma(q)$, then $\varphi^{\#}(p) = \varphi^{\#}(q)$. However, if $p \notin \Gamma q$, then $\varphi^{\#}(p) \perp \varphi^{\#}(q)$.

In view of this theorem, we may speak of $\varphi^{\#}(w)$, $w \in S(r)/\Gamma$.

Theorem 4: $\theta(r) = \oplus \Sigma_{w\in S(r)/\Gamma} \varphi^{\#}(w)$.

We will now see how to compute with these theorems in our special case.

Consider $\exp 2\pi i(nz) \in H_1^{\wedge}$, $n \neq 0$. Then the Y orbit of this character in $H_1^{\wedge}$ consists of the characters of the form $\exp 2\pi i(nm_1 x+nz)$. Hence if $Tf = F$, $f \in L^2(B\backslash N)$ we have

$$AF(\Gamma g) = \Sigma_{x\in\Gamma/L} F(xg)$$

$$= \Sigma_{m\in Z} F((0,m,0)(x,y,z)) = \Sigma_{m\in Z} F(x,y+m,z+mx)$$

$$= \Sigma_{m\in Z} f(y+m)\, e^{2\pi inmx}\, e^{2\pi inz} \quad .$$

Hence if $H_{n,0}$ denotes the space of functions on $\Gamma\backslash N$ of the above form it is in H_n and R-irreducible. We note that for $n = 1$, $H_{1,0} = H_1$ and $AT = \mathcal{J}\Theta$ and we have finally arrived at our desired explaination of why the mapping Θ works. Thus the mapping Θ works because the maximal subgroups of N leaving the character $\exp 2\pi iz$ on B invariant is B itself. This means that the spaces

$$\mathbb{C} \exp 2\pi i(m_1 z+z) \quad \text{and} \quad \mathbb{C} \exp 2\pi i(m_2 x+z) \quad , \; m_1 \neq m_2$$

are inequivalent irreducible subspaces of $L^2(L\backslash B)$ under a unitary representation of B on this space and so are orthogonal.

Let us now return to the case where we consider $\exp 2\pi i(nz) \in \hat{H}_1$, $n \neq 0,\pm 1$. Then the Y orbits of these characters does not exhaust all characters of the form

$$\exp 2\pi i(mx+nz) \quad , \quad m \in Z \quad .$$

To exhaust these characters by Y orbits we must consider orbits through $\exp 2\pi i(\alpha z+nz)$, $0 \leq \alpha < |n|$ also. Each of these orbits determines a direct summand of H_n, $H_{n,\alpha}$, spanned by functions of the form

$$\Sigma_{m\in Z} f(y+m)\, e^{2\pi i(\alpha+mn)x}\, e^{2\pi inz}$$

and

$$H_n = \oplus \sum_{0\leq\alpha<|n|} H_{n,\alpha}$$

and thus H_n is a multiplicity space of multiplicity $|n|$. Further $H_{n,\alpha}$ has orthonormal basis $e^{2\pi i\alpha x}\, e^{2\pi i(anx+by)}\, e^{2\pi inz}$, $a,b \in Z$.

This completes the technical discussion we will require for the rest of these notes. However it may be worthwhile to pause and philosophically discuss why this should all have been expected to work.

We begin by letting $A \subset N$ be the normal abelian subgroup, where $x = 0$. Let $\chi_A(y,z) = \exp(2\pi iz)$. Similarly, let $\chi_B(x,z) = \exp(2\pi iz)$ be a character on B. There are two consequences that one may draw from the classical Stone-von Neumann theorem or a slightly more general version of the Mackey theorems than quoted above.

Let V_A and V_B be the unitary representations of N obtained by inducing χ_A and χ_B from A and B respectively to N, then

1. V_A and V_B are irreducible ;
2. V_A and V_B are unitarily equivalent.

However, we may make (2) more explicit by writing down an intertwining operator between V_A and V_B. To do this, let $\mathbb{R} = N/A$ and $\mathbb{R}^* = N/B$. Then $\mathbb{R}$ and $\mathbb{R}^*$ are the reals as abelian groups with natural parameters x and y, respectively. Further, we may view V_A as unitary operators on $L^2(\mathbb{R})$ and V_B as unitary operators on $L^2(\mathbb{R}^*)$. Let us now view $\mathbb{R}^*$ as the dual group of $\mathbb{R}$ under the pairing

$$(x,y) \to \exp(2\pi ixy) .$$

Then if $f \in L^2(\mathbb{R})$ and $\mathcal{F}(f)$ is the Fourier transform of f defined by

$$\mathcal{F}(f)(-2\pi y) = \lim_{N\to\infty} \int_{-N}^{N} f(x)\, e^{-2\pi ixy}\, dy$$

where the limit is in the $L^2(\mathbb{R}^*)$ topology, one verifies directly that

$$\mathcal{F}V_A = V_B\mathcal{F}$$

and so the Fourier transform is an explicit intertwining operator for the equivalent irreducible representations V_A and V_B.

In the group N again consider the subgroup C of all elements $(0,0,n)$, $n \in Z$, where Z is the integers. Then C is a discrete central subgroup of N and so we may form the quotient group $N^{\#} = N/C$. ($N^{\#}$ is a connected Lie group that has no faithful matrix representation.) We note that V_A and V_B are faithful unitary representations of $N^{\#}$ and further if $\mathbf{T}$ denotes the circle group, we may identify $\mathbf{T}$ with the center of $N^{\#}$ and then $N^{\#}$ satisfies the exact sequence

$$1 \to \mathbf{T} \to N^{\#} \to \mathbb{R} \oplus \mathbb{R}^{*} \to 1 \tag{1}$$

The cocycle defining the group $N^{\#}$ in (1) is given by $(x,y) \to \exp(2\pi ixy)$. Thus we see that the pairing of $\mathbb{R}$ and $\mathbb{R}^*$ to $\mathbf{T}$ that makes $\mathbb{R}^*$ the dual group of $\mathbb{R}$ completely determines the group structure of $N^{\#}$. Therefore it is reasonable to expect the group theory and the harmonic analysis of $N^{\#}$ to reflect the harmonic analysis of $\mathbb{R}$ and vice versa.

The above philosophical statement can be made more precise as follows: We see that $\mathbb{R}$ and $\mathbb{R}^*$ play a more or less symmetric role in $N^{\#}$ and so there should be an automorphism of $N^{\#}$ that interchanges $\mathbb{R}$ and $\mathbb{R}^*$. Since this would interchange the groups A and B it is reasonable to expect that such an automorphism should be related to the Fourier transform $\mathcal{J}$. This is indeed the case, but a little more care is necessary, because V_A and V_B depend not only on A and B, but also on the characters χ_A and χ_B. Thus we are seeking an automorphism of $N^{\#}$ that maps A to B and χ_A to χ_B. This is done by the automorphism $\mathfrak{a}^{\#}$ which was discussed earlier.

We must now find a "natural" L^2 function space on which $\mathfrak{a}^{\#}$ induces a unitary operator and which contains a discrete subrepresentation equivalent to V_A and V_B. This is exactly the role played by $H_1 \subset L^2(\Gamma\backslash N)$.

CHAPTER II

FUNCTIONS ON $\Gamma\backslash N$ AND THETA FUNCTIONS

In Chapter I, we saw that for $f(t) \in L^2(\mathbb{R})$, $\mathcal{J}\Theta f(t) \in H_1$. Perhaps the best understood function in $L^2(\mathbb{R})$ that is not of compact support is the function $e^{-\pi t^2}$, which is sometimes called the Gauss kernel or the normal distribution. It seems reasonable to try and explicitly compute the function $\mathcal{J}\Theta\, e^{-\pi t^2}$. This is what we will now do.

$$\mathcal{J}\Theta\, e^{-\pi t^2} = e^{2\pi i z} \sum_{m\in Z} e^{2\pi i m x}\, e^{-\pi(m+y)^2}$$

which by elementary operations of expanding and rewriting may be written in the form

$$\mathcal{J}\Theta\, e^{-\pi t^2} = e^{2\pi i z}\, e^{-\pi y^2} \sum_{m\in Z} \exp \pi i(im^2+2m\zeta)$$

where $\zeta = x + iy$. But now the sum on the right is a classical Jacobi theta function. This computation suggests that there may exist a deep relationship between the functions in the function spaces $H_n \subset L^2(\Gamma\backslash N)$ and the classical Jacobi theta functions. In this chapter we will see that this is indeed the case.

Let us begin by defining the Jacobi theta functions and a class of functions that include the Jacobi theta functions. To begin with let $n \in Z^+$, the positive integers, let $a, b \in \mathbb{R}$ and let $\mathcal{H}$ denote the upper half plane in $\mathbb{C}$, i.e., $\tau \in \mathcal{H}$ if and only if $\mathrm{Im}(\tau) > 0$. We define a generalized n^{th}-order theta function of period τ and characteristic $[{a \atop b}]$ as a continuous function $G(\zeta)$, $\zeta = x+iy$ on $\mathbb{C}$ satisfying the following functional equations:

1) $G(\zeta+1) = (-1)^a\, G(\zeta)$, $\quad \zeta \in \mathbb{C}$

2) $G(\zeta+\tau) = (-1)^b \exp(-\pi i n(2\zeta+\tau))\, G(\zeta)$, $\quad \zeta \in \mathbb{C}$

The space of all n^{th}-order generalized theta functions of period τ and characteristic $[{a \atop b}]$ will be denoted by $\Theta_n[{a \atop b}](\tau)$. In $\Theta_n[{a \atop b}](\tau)$ we may distinguish those functions which are entire functions of ζ. These entire functions are the classical Jacobi theta functions and will be denoted by $\Theta_n[{a \atop b}](\tau, A)$.

In order to state our next result we must define some function spaces on $\Gamma\backslash N$. Let $C^k(\Gamma\backslash N)$, $0 \le k \le \infty$, be the functions on $\Gamma\backslash N$ of class k with the convention that $C^0(\Gamma\backslash N)$ is the space of continuous functions on $\Gamma\backslash N$. We next define $C_n^k = C^k(\Gamma\backslash N) \cap H_n$. Since the projection operator $p_n: L^2(\Gamma\backslash N) \to H_n$ has the property of preserving the function spaces $C^k(\Gamma\backslash N)$, $0 \le k \le \infty$, it follows that C_n^k is dense in H_n for all n and k. We will also often adopt the notation convention that $C_n^0 = C_n$.

The example that we discussed at the beginning of this chapter suggests that we try to prove that $e^{2\pi inz} e^{-\pi ny^2} \Theta_n[{a \atop b}](\tau) = C_n$. This is not quite right. A small amount of computational experimentation suggests the correct result as stated below as Theorem II.1.

Theorem II.1: Let $n \in Z^+$, $a,b \in \mathbb{R}$, and $\tau \in \mathcal{H}$. Set

$$(3) \quad \left\{M_n[{a \atop b}](\tau)(G)\right\}(x,y,z) = e^{2\pi inz}\, e^{\pi i\tau ny^2}\, e^{2\pi in(\frac{b}{2}y - \frac{a}{2}x)}\, G(x+\tau y)$$

where $G \in \Theta_n[{a \atop b}](\tau)$ and $(x,y,z) \in N$. Then $M_n[{a \atop b}](\tau)(G) \in C_n$ and $M_n[{a \atop b}](\tau)$ defines a $\mathbb{C}$-linear isomorphism of $\Theta_n[{a \atop b}](\tau)$ and C_n.

Proof: For $G \in \Theta_n[{a \atop b}](\tau)$ put

$$F(x,y,z) = (M_n[{a \atop b}](\tau)(G))(x,y,z) .$$

We must show that $F(x,y,z) \in C_n$. It is straightforward to verify that $F(x,y,z+t) = e^{2\pi int}F(x,y,z)$. Next we must verify that the transformation $(n_1,n_2,n_3)(x,y,z) \to (n_1+x, n_2+y, n_2x+n_3+z)$ leaves F invariant, where $(n_1,n_2,n_3) \in \Gamma$. Because $(1,0,0)$ and $(0,1,0)$ generate Γ, we must only show F is invariant under these two transformations. Since

$$(1,0,0)(x,y,z) = (x+1,0,0)$$

(1) and (3) easily combine to yield the desired invariance. Now the fact that $(0,1,0)(x,y,z) = (x,y+1,x+z)$ implies that

$$F((0,1,0)(x,y,z)) = e^{2\pi i(\frac{b}{2}(y+1) - \frac{a}{2}x)}\, e^{2\pi i(x+z)}\, e^{2\pi i\tau x(y+1)^2}\, G(x+\tau y+\tau)$$

$$= e^{2\pi i \frac{b}{2}}\, e^{\pi i n(2(x+\tau y)+\tau)}\, e^{2\pi i((\frac{b}{2}y)-(\frac{a}{2})x+nz))}\, e^{\pi i\tau m y^2}\, G(x+\tau y+\tau)$$

$$= F(x,y,z)$$

using condition (2). Thus $M_n[{a \atop b}](\tau)(G) \in C_n$ whenever $G \in \Theta_n[{a \atop b}](\tau)$.

The converse may be obtained by reversing the steps in the computation and so we have proven our result.

It is worthwhile at this junctor to pause to examine two important algebraic properties of the mappings $M_n[{a \atop b}](\tau)$. It is trivial to verify that if $F_n \in C_n$ and $F_k \in C_k$ then $F_n \cdot F_k \in C_{n+k}$. Similarly, if $G_n \in \Theta_n[{a \atop b}](\tau)$ and $G_k \in \Theta_k[{a \atop b}](\tau)$, $n,k \in Z^+$ then $G_n \cdot G_k \in \Theta_{n+k}[{a \atop b}](\tau)$. It is then straightforward to verify that the diagram below is commutative

$$\begin{array}{ccc}
\Theta_n[{a \atop b}](\tau) \times \Theta_k[{a \atop b}](\tau) & \longrightarrow & \Theta_{k+k}[{a \atop b}](\tau) \\
\Big\downarrow M_n[{a \atop b}](\tau) \times M_k[{a \atop b}](\tau) & & \Big\downarrow M_{n+k}[{a \atop b}]\ \tau \\
C_n \times C_k & \longrightarrow & C_{n+k}
\end{array}$$

and horizontal arrows denote multiplication of functions.

Let us now consider C_0 . These are the functions on $\Gamma\backslash N$, that are invariant under the action of the group $(0,0,z)$, $z \in \mathbb{R}$. It is easy to verify that $\Gamma\backslash N_3$ is a principal circle bundle over the torus $\mathbf{T}^2$ with the orbits of $(0,0.z)$ on $\Gamma\backslash N_3$ as fibers. If $\pi\colon \Gamma\backslash N_3 \to \mathbf{T}^2$ denotes the fiber mapping then $F \in C_0$ if and only if $F = E \circ \pi$, where E is a function on $\mathbf{T}^2$. It is straightforward to verify that if $g \in C_0$ if $G \in \Theta_n[{a \atop b}](\tau)$ and $F \in C_n$ then $gG \in \Theta_n[{a \atop b}](\tau)$, $gF \in C_n$ and

$$M_n[{a \atop b}](\tau)(gG) = g\cdot(M_n[{a \atop b}](\tau)(G)) \quad .$$

In view of Theorem II.1 it is reasonable to ask which properties of Jacobi theta functions hold for functions in $\mathcal{C}_n$, $n \in \mathbf{Z} - \{0\}$, and, further, which properties of Jacobi theta functions hold for generalized theta functions.

Before beginning to look directly at this problem let us note that the mapping $\mathfrak{J}: (x,y,z) \to (x_1-y,-z)$ is an automorphism of N that maps Γ onto itself and so $\mathfrak{J}^*$ is a homeomorphism of $\Gamma\backslash N$ onto itself that induces a unitary operator $\mathfrak{J}^*_U$ on $L^2(\Gamma\backslash N)$ with the property that $\mathfrak{J}^*_U$ is an intertwining operator between H_n and H_{-n}, $n \neq 0$. This enables us to prove theorems for $n \in \mathbf{Z}^+$ and conclude that they hold for $n \in \mathbf{Z} - \{0\}$.

It is a classical result that if $G \in \Theta_n[{a \atop b}](\tau,A)$ then for some ζ_0 in the unit square $G(\zeta_0) = 0$, where ζ_0 is in the unit square if $\zeta_0 = x_0 + iy_0$ implies $0 \le x_0, y_0 < 1$. We will see that this holds if $G \in \Theta[{a \atop b}](\tau)$ and if $F \in \mathcal{C}_n$, $n \neq 0$, then $F(x_0,y_0,z) = 0$ for some x_0,y_0 and all $z \in \mathbb{R}$. Before proving this assertion we will have to discuss a little more fiber bundle structure.

Let $\Gamma(n)$ be the subgroup of N_3 consisting of the groups generated by Γ and the element $(0,0,1/n)$. Then

$$\begin{array}{c}\Gamma\backslash N\\ \downarrow \pi(n)\\ \Gamma(n)\backslash N\end{array}$$

is a covering bundle. Now let $F \in H_n$. Then

$$F(x,y,z+\frac{1}{n}) = e^{2\pi i(1/n)n}\, F(x,y,z) = F(x,y,z) .$$

Hence F may be viewed as a function on $\Gamma(n)\backslash N$. Clearly $\Gamma(n)\backslash N$ is also a principal fiber bundle over $\mathbf{T}^2$.

<u>Theorem II.2</u>: If $F \in \mathcal{C}_n$, $n \neq 0$, then F vanishes on some fiber of the bundle $\pi: \Gamma\backslash N \to \mathbf{T}^2$.

<u>Proof</u>: Assume the theorem is false. Then $F \in \mathcal{C}_n$, $F \neq 0$. Now $|F|$ is constant on fibers of π and so is in $\mathcal{C}_0$ and hence

$$K = F/|F| \in \mathcal{C}_n .$$

Since $K \in C_n$, K is a continuous function on $\Gamma(n)\backslash N$. Hence the set of points $X \subset \Gamma(n)\backslash N$ where K assumes the value one is a closed subset of $\Gamma(n)\backslash N$. We will now prove that K is a cross section over T^2 for the principal bundle,

$$\begin{array}{c}\Gamma(n)\backslash N\\ \downarrow\\ T^2\end{array}$$

The existence of K would imply $\Gamma(n)\backslash N$ is a product bundle and so a three dimensional torus. But this is absurd since the fundamental group of $\Gamma(n)\backslash N$ is $\Gamma(n)$ and $\Gamma(n)$ is non-abelian. This will complete our proof.

Now let $\mathcal{J}$ be a fiber of $\Gamma(n)/N_3$ over T^2. Then K takes on the values $K(x_0,y_0,z_0)\, e^{2\pi i n t}$, $0 \le t < 1/n$, where $x_0,y_0,z_0 \in \mathcal{J}$. Hence K assumes the value 1 exactly once on each fiber $\mathcal{J}$. Hence X is a cross section of the bundle $\Gamma(n)\backslash N$ over T^2 which is impossible.

<u>Corollary II.3</u>: Let $G \in \Theta_n[{a \atop b}](\tau)$, $n \in Z^+$, then there exists $\zeta_0 = x_0 + iy_0$ where $0 \le x_0, y_0 < 1$ such that $G(\zeta_0) = 0$.

This follows easily from Theorems II.2 and II.1 and the fact that $e^{2\pi i n z}\, e^{\pi i \tau n y^2}\, e^{2\pi i n((b/2)y-(a/2)x)} \ne 0$.

For the rest of this section we will use facts about the zeros of the Jacobi theta functions to obtain some fairly deep insight into the spaces C_n^k and H_n and their inter-relation. Thus for much of the remainder of this chapter we will concentrate our attention on $\Theta_n[{0 \atop 0}](i)$ and $\Theta_n[{0 \atop 0}](i,A)$. To simplify notation for the rest of this chapter we will denote $\Theta_n[{0 \atop 0}](i)$ and $\Theta_n[{0 \atop 0}](i,A)$ by Θ_n and $\Theta_n(A)$, respectively, and $M_n[{0 \atop 0}](i)$ by M_n.

We begin by noting that an elementary computation will verify that the function

$$\theta[{a \atop b}](\zeta,\tau) = \Sigma_{m\in Z} \exp \pi i(\tau(m+(a/2))^2 + 2(m+(a/2))(\zeta+(b/2)))$$

is in $\Theta_1[{a \atop b}](\tau)$. We next define

$$\theta_{n\nu}(\zeta) = \theta\begin{bmatrix}\frac{2\nu}{n}\\ 0\end{bmatrix}(n\zeta, ni) \quad , \quad \nu = 0,\ldots,n-1 .$$

It is well known and easily verified that $\theta_{n\nu}(\zeta) \in \Theta_n(A)$, $\nu = 0,\dots,n-1$. We will also denote $M_n(\theta_{n\nu})$ by $\psi_{n\nu}$ and define

$$\psi^*_{n\nu} = R(\tfrac{1}{2},\tfrac{1}{2},\tfrac{1}{4})\psi_{n\nu} \ , \quad \nu = 0,\dots,n-1 \ .$$

Before we can state our next two main results let $C^k(T^2;(1/n))$ denote the doubly periodic functions $f(x,y)$ in the plane of class k such that

$$f(x + (1/n),y) = f(x,y)$$

$$f(x,y+1) = f(x,y) \ .$$

Let us now state the two results whose proofs will be completed in several stages.

<u>Theorem II.4</u>: Let $F \in C^k_n$, $n \in Z^+$. Then there exists $g_\nu, h_\nu \in C^k(T^2,(1/n))$, $\nu = 0,\dots,n-1$ such that

$$F = \sum_{\nu=0}^{n-1} (g_\nu \psi_{n\nu} + h_\nu \psi^*_{n\nu}) \ .$$

Let $F \in C^k_n$, $n \in Z^-$, where Z^- is the negative integers, then there exists $g_\nu, h_\nu \in C^k(T^2,(1/n))$, $\nu = 0,\dots,n-1$ such that

$$F = \sum_{\nu=0}^{n-1} (g_\nu \mathfrak{J}^*_U(\psi_{n\nu}) + h_\nu \mathfrak{J}^*_U(\psi^*_{n\nu})) \ .$$

<u>Theorem II.5</u>: The set of functions $\psi_{n\nu} C^\infty(T^2,(1/n))$, $\nu = 0,\dots,n-1$ is dense in H_n, $n \in Z^+$ in the L^2 norm. The set of functions $\mathfrak{J}^*_U(\psi_{n\nu}) C^\infty(T^2,(1/n))$ is dense in H_n, $n \in Z^-$.

It is clear from our discussion of the mapping $\mathfrak{J}$ that once we have proven these two theorems for $n \in Z^+$, the proofs will follow immediately for $n \in Z^-$. We will now prove Theorems II.4 and II.5 in the special case where $n = 1$ and then proceed to see how to reduce the general case to this special case.

It is well known that the function $\theta_{10}(\zeta) = 0$, $\zeta = x + iy$ only for $x = y = \tfrac{1}{2}$ provided $0 \le x,y < 1$. Hence the function ψ_{10} is zero only on the image of $(\tfrac{1}{2},\tfrac{1}{2},z)$ in $\Gamma\backslash N$. Since ψ^*_{10} is a translate of ψ_{10} by an element of N , it easily follows that ψ^*_{10} is zero only on the image $(0,0,z)$ in $\Gamma\backslash N_3$

Now in $\mathbf{T}^2 = \mathbb{R} \times \mathbb{R}/\mathbf{Z} \times \mathbf{Z}$ consider

$$B^* = \{(x,y) \mid \epsilon < y < 1 - \epsilon \ , \ 0 \leq x < 1\}$$

and $B = B^* + (\frac{1}{2},\frac{1}{2})$, where addition is in $\mathbf{T}^2$.

Let $\pi\colon \Gamma\backslash N \to \mathbf{T}^2$ be the fiber bundle projection and let $X = \pi^{-1}(B)$ and $X^* = \pi^{-1}(B^*)$. Then X and X^* retract onto a two dimensional torus and so each of the bundles X and X^* have a cross-section which we will denote by S and S^*, respectively. Further $\psi_{10}|S$ and $\psi^*_{10}|S^*$ are bounded away from zero. Now let $F_1, F_2 \in C^k_1$ be such that $F_1|S = F_2|S$. Then $F_1 = F_2$ in $\pi^{-1}(B)$ and a similar statement holds with stars.

Again let $F \in C^k_1$ and let $F|S = F$ and $F|S^* = F^*$. Then, since $\psi_{10}|S$ is bounded away from zero, there exists k and $k^* \in C^k(\mathbf{T}^2,1)$ such that

$$k\, \psi_{10}|S = F$$
$$k^* \psi^*_{10}|S^* = F^* \quad .$$

Observe that B and B^* is an open covering of $\mathbf{T}^2$ and so we may choose a C^k partition of unity ξ_α, $\alpha = 1,\ldots,N$ with the property that if $C(\xi_\alpha)$ denotes the carrier of ξ_α, then

1) $C(\xi_1),\ldots,C(\xi_k) \subset B$

2) $C(\xi_{k+1}),\ldots,C(\xi_N) \subset B^*$.

Now

$$F = \sum_{\alpha=1}^{N} F\xi_\alpha$$

and $C(g\ \xi_\alpha) \subset B$ for $\alpha = 1,\ldots,k$ and $C(F\ \xi_\alpha) \subset B^*$ for $\alpha = k+1,\ldots,N$. Hence there exists $k_\alpha \in C^k(\mathbf{T}^2,1)$ such that $C(k_\alpha) = C(\xi_\alpha)$ and

$$F\ \xi_\alpha|S = k_\alpha\ \psi_{10}|S \ , \quad \alpha = 1,\ldots,k$$
$$F\ \xi_\alpha|S^* = k_\alpha\ \psi^*_{10}|S^* \ , \quad \alpha = k+1,\ldots,N \ .$$

Let $g_1 = \Sigma^k_{\alpha=1} k_\alpha$ and $h_1 = \Sigma^N_{\alpha=k+1} k_\alpha$ then since $k_\alpha \in C^k(\mathbf{T}^2,1)$, g_1 and h_1 are also in $C^k(\mathbf{T}^2,1)$ and so we have proven Theorem II.4 in the special case where $n = 1$ or $F = g_1\ \psi_{10} + h_1\ \psi^*_{10}$.

Let us now pause to make an important observation. Let $F \in C_1^k$ and assume $\lim \frac{F}{\psi_{10}}$ exists at all points of $\Gamma\backslash N$. We will now see how the above proof can be modified to show that

$$F = g_1 \psi_{10} \quad , \quad g_1 \in C^0(\mathbf{T}^2,1)$$

Choose a C^k partition of unity ξ_{α_1}, $\alpha = 1,\ldots,N$ for $\mathbf{T}^2$ such that $C(\xi_\alpha)$, $\alpha = 1,\ldots,N$ is contractable. Then there exists a cross-section S_α over $C(\xi_\alpha)$, $\alpha = 1,\ldots,N$. Choose $h_\alpha \in C^0(\mathbf{T}^2,1)$ such that

$$F\xi_\alpha = h_\alpha \xi_\alpha \psi_{10} \quad , \quad \alpha = 1,\ldots,N .$$

This can be done because we have assumed $\lim \frac{g}{\psi_{10}}$ exists. Then

$$F = \Sigma F\xi_\alpha = \psi_{10} \Sigma h_\alpha \xi_\alpha = g_1 \psi_{10}$$

where $g_1 \in C^0(\mathbf{T}^2,1)$.

This has an important consequence for us. If $F \in C_1^0$ vanishes in a neighborhood of $(\frac{1}{2},\frac{1}{2},z)$ in $\Gamma\backslash N$ it is in $\psi_{10} C_1^0(\mathbf{T}^2,1)$.

The above result has as an almost immediate corollary the special case of Theorem II.5 where $n = 1$: i.e., the set of functions $\psi_{10} C^0(\mathbf{T}^2,1)$ is dense in H_1 in the L^2 norm.

We begin by recalling that C_1^0 is dense in H_1 in the L^2 norm. Now let $F \in C_1^0$. Then we may easily find F_β, $\beta = 1,2,\ldots,$ such that $F_\beta \in C_1^0$, F_β vanishes in a neighborhood of $(\frac{1}{2},\frac{1}{2},z)$ and the limit of the F_β in the L^2 norm is F. But $F_\beta = g_\beta \psi_{10}$, $g_\beta \in C^0(\mathbf{T}^2,1)$, $\beta = 1,2,\ldots$. This proves that $\psi_{10} C^0(\mathbf{T}^2,1)$ is dense in H_1 in the L^2 norm.

Let us now consider the mapping

$$I : L^2(\mathbf{T}^2,1) \to H_1$$

given by $I: g \to \psi_{10}g$, $g \in L^2(\mathbf{T}^2,1)$. This mapping is clearly continuous and 1-1, since $\psi_{10}g = 0$, a.e. implies $g = 0$, a.e. . Hence, by the open mapping theorem, we have that I^{-1} is also continuous. It follows that if φ_β is an orthonormal basis of $L^2(\mathbf{T}^2,1)$ then $I(\varphi_\beta)$ has the following two properties:

3) $I(\varphi_\beta)$, $\beta = 1,2,\ldots$ is dense in H_1 .

4) No proper subset of $I(\varphi_\beta)$, $\beta = 1,2,\ldots$ is dense in H_1 .

We will call a set of functions in a Hilbert space with the above two properties in L^2 basis of the Hilbert space or a Hilbert space basis.

In this language we may restate our result as follows:

$$\psi_{10} \exp 2\pi i(ax+by), \quad a,b \in \mathbf{Z} \text{ is an } L^2 \text{ basis of } H_1 .$$

Note the above functions are in $C^\infty(\Gamma\backslash N)$, but are not orthonormal.

Before we can consider the general case of Theorems II.4 and Theorems II.5 we must develop some of the ideas first presented in a joint work of L. Auslander and J. Brezin [1]. As remarked earlier, if $F \in H_n$ we may view $F \in L^2(\Gamma(n)\backslash N)$, where $\Gamma(n)$ is the subgroup of N consisting of elements of the form $(m_1, m_2, (m_3/n))$, $m_i \in \mathbf{Z}$, $i = 1,2,3$. Now let $\Lambda(n)$ be the subgroup of N given by the elements of the form $(m_1/n, m_2/n, m_3/n^2)$, $m_i \in \mathbf{Z}$, $i = 1,2,3$. Note that $\Gamma \subset \Gamma(n) \subset \Lambda(n)$ and that $\Gamma(n)$ is normal in $\Lambda(n)$. For the time being, we will view H_n as functions on $\Gamma(n)\backslash N$. On the Hilbert space $L^2(\Gamma(n)\backslash N)$, we shall define two representations; one of N, the other by $\Lambda(n)/\Gamma(n)$ by

$$(R(g)F)(\Gamma(n)h) = F(\Gamma(n)hg) \quad , \quad h,g \in N,\ F \in L^2(\Gamma(n)\backslash N)$$

$$(L(\lambda)F)(\Gamma(n)h) = F(\Gamma(n)\lambda^{-1}h) \ , \ \lambda \in \Lambda(n) .$$

Because $\Gamma(n)$ is in the kernel of L, we will view L as a representation of $\Lambda(n)/\Gamma(n)$. Further, it is clear that

$$L(\lambda)\, R(g) = R(g)\, L(\lambda) \ , \quad g \in N \text{ and } \lambda \in \Lambda(n) .$$

In Chapter I we outlined a proof of the assertion that H_n is a multiplicity space of multiplicity $|n|$. Thus if we fix an irreducible R-invariant subspace $\mathcal{J}$ in H_n , there is an isometric isomorphism V from H_n to $\mathcal{J} \otimes \mathbb{C}^n$ such that

$$(R(g) \otimes 1)V(F) = VR(g)F \quad , \quad g \in N \text{ and } F \in H_n .$$

Proposition 1 of [1] then asserts that there is a unitary representation L^n of $\Lambda(n)/\Gamma(n)$ on $\mathbb{C}^n$ such that for all $\lambda \in \Lambda(n)/\Gamma(n)$ and all $F \in H_n$ we have

$$VL(\lambda)F = (1 \otimes L^n)VF \quad .$$

Furthermore, L^n does not depend, up to unitary equivalence on the choice of $\mathcal{J}$ or V .

Let $z(\Lambda(n)/\Gamma(n))$ denote the center of $\Lambda(n)/\Gamma(n)$ which is given by the cosets $\Gamma(n)(0,0,a/n^2)$, $a \in \mathbf{Z}$. Further, let η denote the character on $z(\Lambda(n)/\Gamma(n))$ given by

$$\Gamma(n)(0,0,a/n^2) \to \exp(2\pi i\ a/n) \ .$$

Clearly the restriction of L from $\Lambda(n)/\Gamma(n)$ to $z(\Lambda(n)/\Gamma(n))$ is a multiple of η and hence the restriction of L^n to $z(\Lambda(n)/\Gamma(n))$ is a multiple of η . It is classical, but a proof is given in [1] that there is, up to unitary equivalence, precisely one irreducible representation $T(n)$ of $\Lambda(n)/\Gamma(n)$ whose restriction to $z(\Lambda(n)/\Gamma(n))$ is a multiple of η . The dimension of $T(n)$ is $|n|$, and further, L^n is unitarily equivalent to $T(n)$.

It may be worthwhile to give another model of $\Lambda(n)/\Gamma(n)$ and a matrix representation of $T(n)$. Let

$$\Gamma = \begin{pmatrix} 1 & m_2 & m_3 \\ 0 & 1 & m_1 \\ 0 & 0 & 1 \end{pmatrix} \qquad m_i \in \mathbf{Z} \ , \ i = 1,2,3 \quad .$$

and look at the group obtained by replacing m_i by m_i mod n, $i = 1,2,3$. This group is isomorphic to the group $\Lambda(n)/\Gamma(n)$.

Let $e_0,\ldots,e_{n-1}$ be a basis of $\mathbf{C}^n$ and let A be the linear transformation of $\mathbf{C}^n$ defined by

$$A(e_j) = \exp(2\pi i\ j/n)\ e_j \quad , \quad j = 0,\ldots,n-1$$

and let B be the linear transformation defined by

$$B(e_j) = e_{j+1} \ , \quad j = 0,\ldots,n-1$$

where addition in the subscript is taken mod n. We see easily that $A^n = I$ and $B^n = I$, and

$$B^{-1}A^{-1}BA(e_j) = \exp(2\pi i/n)\, e_j \ , \quad j = 0,\dots,n-1 \ .$$

Thus the group generated by the matrices A and B is isomorphic to $\Lambda(n)/\Gamma(n)$ and the matrix representation thus given is unitarily equivalent to $T(n)$.

Consider now the functions $\psi_{u\nu}$ and $\theta_{u\nu}$, $\nu = 0,\dots,n-1$ defined earlier in this chapter. Now

$$L(\tfrac{a}{n},0,0) = e^{2\pi inz}\, e^{-\pi ny^2} \sum_m \exp \pi i\{ni(m+\tfrac{\nu}{n})^2 + 2n(m+\tfrac{\nu}{n})iy\}\cdot\exp 2\pi in(m+\tfrac{\nu}{n})(x+\tfrac{a}{n})$$

$$= \exp(2\pi ia\,\tfrac{\nu}{n})\psi_{n\nu} \ , \quad \nu = 0,\dots,n-1 \ .$$

A similar computation shows that

$$L(0,\tfrac{1}{n},0)\psi_{n\nu} = \psi_{n(\nu+1)} \ , \quad \nu = 0,\dots,n-1$$

where the addition in the subscript is mod n. Thus we have an irreducible representation of $\Lambda(n)/\Gamma(n)$ on the vector space spanned by $\psi_{n\nu}$ in H_n. These computations may be summarized as follows:

<u>Theorem II.6</u>: Let $H_{n\nu} \subset H_n$ be the eigenfunction space of the operator $L(\frac{1}{n},0,0)$ with eigenvalue $\exp(2\pi i\,\frac{\nu}{n})$, $\nu = 0,\dots,n-1$. Then $H_{n\nu}$ is an irreducible subspace for the unitary representation R of N restricted to H_n and

1) $H_n = \oplus \sum_{\nu=0}^{n-1} H_{n\nu}$, where the sum is an orthogonal direct sum.

2) $\psi_{n\nu} \in H_{n\nu}$.

3) $L(0,\frac{\nu}{n},0)$ is an intertwining operator between H_{n0} and $H_{n\nu}$ and $L(0,\frac{\nu}{n},0)\psi_{n0} = \psi_{n\nu}$, $\nu = 0,\dots,n-1$

4) H_{n0} may be identified with the H_1 space to $\Gamma^*\backslash N$ where $\Gamma^* = (m_1/n, m_2, m_3/n)$, $m_i \in Z$, $i = 1,2,3$.

Let us begin by stating two further results.

<u>Theorem II.7</u>: Let $M_n^{-1}\, L(\frac{a}{n},0,0)M_n = L^*(\frac{a}{n})$ operate on Θ_n. Then

1) $L^*(\frac{a}{n})\, \Theta_n(A) = \Theta_n(A)$

2) If $\theta \in \Theta_n(A)$ and $p_\nu(\theta)$ is the projection of θ onto the eigen function with eigen value $\exp(2\pi i \frac{\nu}{n})$, then $p_\nu(\theta) \in \Theta_n(A)$.

Theorem II.8: Let $p_\nu: H_n \to H_{n\nu}$ be the orthogonal projection. If $F \in C_n^k$, then $p_\nu(F) \in C_n^k$, all k . Further, if $F(x_0,y_0,z) = 0$ for $(x_0,y_0,z) \in \Gamma\backslash N$, then $F(x_0+ \frac{1}{n},y_0,z) = 0$.

We will prove only Theorem II.7 as the proof of Theorem II.8 then follows by essentially the same arguments.

Proof of Theorem II.7: Let $\theta \in \Theta_n(A)$. Then $L^*(\frac{a}{n})\theta(\zeta) = \theta(\zeta+ \frac{a}{n})$. As can be seen by a direct and elementary argument. Thus our major concern must be with the proof of assertion (2). Now consider $\Theta_n(A) \subset \Theta_n$. Then $\Theta_n = \oplus \Sigma_{\nu=0}^{n-1} \Theta_{n\nu}$, where $\Theta_{n\nu}$ consists of the eigenfunctions of $L^*(\frac{1}{n})$ with eigenvalue $\exp(2\pi i \frac{\nu}{n})$, $\nu = 0,\ldots,n-1$. For $\theta \in \Theta_n(A)$, we must show that $p_\nu(\theta) \in \Theta_n(A)$. Now

$$p_0(\theta) = \frac{1}{n}(\theta + L^*(\frac{1}{n})\theta +\ldots+ L^*(\frac{1}{n})^{n-1} \theta)$$

and so $p_0(\theta) \in \Theta_n(A)$ as is $\theta - p_0(\theta)$. Assume now that $\theta \in \Theta_n(A)$ and $p_0(\theta) = 0$. It is then easily verified that

$$p_{\nu-1}(\theta) = \frac{1}{n}(\theta + \exp \frac{2\pi i}{n} L^*(\frac{1}{n})\theta +\ldots+ \exp \frac{2\pi i(n-1)}{n} L^*(\frac{1}{n})^{n-1} \theta)$$

which proves that $p_{\nu-1}(\theta) \in \Theta_n(A)$. A simple induction then completes the proof of Theorem II.7.

We are now in a position to complete the proof of Theorem II.4.

Proof of Theorem II.4: By Theorem II.8 if we define $C_{n\nu}^k = C^k \cap H_{n\nu}$ then

$$C_n^k = \oplus \sum_{\nu=0}^{n-1} C_{n\nu}^k .$$

By the special case of Theorem II.8 where $n = 1$ already proven and Theorem II.6, we have that

$$C_{n0}^k = \psi_{n0} C_n^k(T^2,\frac{1}{n}) + \psi_{n0}^* C_n^k(T^2,\frac{1}{n}) .$$

Since $L(0,\frac{\nu}{n},0)\psi_{n0} = \psi_{n\nu}$, $L(0,\frac{\nu}{n},0)\psi_{n0}^* = \psi_{n\nu}^*$ and

$$L(0,\frac{\nu}{n},0)C^k_{n0} = C^k_{n\nu}$$

we have easily that

$$C^k_{n\nu} = \psi_{n\nu}\, C^k_n(\mathbf{T}^2,\frac{1}{n}) + \psi^*_{n\nu}\, C^k_n(\mathbf{T}^2,\frac{1}{n})$$

thus

$$C^k_n = \sum_{\nu=0}^{n-1} \psi_{n\nu}\, C^k_n(\mathbf{T}^2,\frac{1}{n}) + \sum_{\nu=0}^{n-1} \psi^*_{n\nu}\, C^k_n(\mathbf{T}^2,\frac{1}{n})$$

and Theorem II.4 is established.

The proof of Theorem II.5 is carried out in an analogous manner by replacing C's with H's and noting that $L(0,\frac{\nu}{n},0)$, $\nu = 1,\ldots,n-1$ are intertwining operators for H_{n0} and $H_{n\nu}$ for the representation R of N .

Thus we have that C^k_n is a module over C^k_0 and we have a set of generators for this module. Before going on to prove our final analytic result, a new proof of a classical result, let us pause to prove the following algebraic result which was shown to us by J. Brezin, but which is essentially a technique of a proof due to Swan.

Theorem II.9: C^k_n is a projective C^k_0 module.

Proof: We observe that the techniques used in the proof of Theorem II.4 apply directly to reduce this theorem to the special case $n = 1$. Thus it remains to prove that C^k_1 is a projective C_0 module.

Let $Q_\epsilon = \{(x,y,0) \in N | 0 \le x < 1,\ \epsilon \le y \le 1-\epsilon\}$ and let

$Q^*_\epsilon = \{(x,y,0) \in N | 0 \le x \le 1,\ 0 \le y \le \frac{1}{2}-\epsilon$ or $\frac{1}{2}+\epsilon \le y \le 1\}$. Let δ,δ^* be a continuous partition of unity for $\mathbf{T}^2$ subordinate to the covering defined by Q_ϵ and Q^*_ϵ . Define

$$\Delta(x,y,z) = \sqrt{\delta(x,y,0)}\quad e^{2\pi i z}$$

$$\Delta^*(x,y,z) = \begin{cases} \sqrt{\delta^*(x,y,0)}\quad e^{2\pi i(z-x)} & \text{if } 0 \le y \le \frac{1}{2} - \epsilon \\ \sqrt{\delta^*(x,y,0)}\quad e^{2\pi i z} & \text{if } \frac{1}{2} + \epsilon \le y \le 1 \end{cases}$$

Then Δ and Δ^* belong to C_1 . Further, if $\bar{z}$ denotes the complex conjugate of $z \in \mathbb{C}$, we have

$$\Delta\bar{\Delta} + \Delta^*(\Delta^*)^- = 1 \, .$$

Hence given $F \in \mathcal{C}_1$, we have

$$\Delta(\bar{\Delta}F) + \Delta^*(\Delta^*)^- F = F \, .$$

Define $\varphi\colon \mathcal{C}_1 \to \mathcal{C}_0$ by $\varphi F = \bar{\Delta}F$ and $\varphi^*\colon \mathcal{C}_1 \to \mathcal{C}_0$ by $\varphi^* F = (\Delta^*)^- F$. Notice that

$$\varphi, \varphi^* \in \mathrm{Hom}_{\mathcal{C}_0}(\mathcal{C}_1, \mathcal{C}_0) \, .$$

Now let $\mathcal{J} = \mathcal{C}_0 e_1 + \mathcal{C}_0 e_2$ be the free $\mathcal{C}_0$ module with basis e_1 and e_2 and let $\sigma\colon \mathcal{J} \to \mathcal{C}_1$ be defined by $\sigma(g_1 e_1 + g_2 e_2) = g_1\Delta + g_2\Delta^*$, $g_1, g_2 \in \mathcal{C}_0$. Then defining

$$\tau\colon \mathcal{C}_1 \to \mathcal{J}$$

by

$$\tau(F) = \varphi(F)e_1 + \varphi^*(F)e_2$$

we have $\sigma \circ \tau = 1$. Hence

$$0 \to \mathrm{Ker}\,\sigma \to \mathcal{J} \to \mathcal{C}_1 \to 0$$

splits and $\mathcal{C}_1$ is projective.

Theorem II.9 raises the interesting question of whether or not $\mathcal{C}(\Gamma\backslash N_3)$ is projective over $\mathcal{C}_0$.

Let us now prove the classical result about the dimension of the vector space of nth order Jacobi theta functions by our approach.

<u>Theorem II.10</u>: The dimension of $\Theta_n(A) = n$, $n \in \mathbf{Z}^+$.

<u>Proof</u>: Let $\xi \in \Theta_n(A)$. We have seen that $\xi = \Sigma_{\nu=0}^{n-1} \xi_\nu$, $\xi_\nu \in \Theta_n(A)$ and further $M_n(\xi_\nu) \in H_{n\nu}$. Now by the argument on page 80-81 of [6], we have that $M_n(\xi_\nu)$ vanishes exactly where $\psi_{n\nu}$ vanishes, $\nu = 0,\ldots,n-1$, and

$$\lim M_n\left(\frac{\xi_\nu}{\psi_{n\nu}}\right)$$

exists in $\Gamma\backslash N$. It follows that

$$\xi_\nu = \theta_{n\nu} f_\nu \, , \qquad f_\nu \in C^0(\mathbf{T}^2, \tfrac{1}{n}) \, .$$

But since f is then bounded in the complex plane and ξ_ν and θ_ν are analytic, we have f_ν is a constant. Hence

$$\xi = \sum_{\nu=0}^{n-1} c_\nu \theta_\nu$$

and we have our assertion.

CHAPTER III

ELEMENTARY TRANSFORMATION THEORY

We have shown that the spaces $H_{n\nu}$, $n \in Z^+$, $\nu = 0,\dots,n-1$ in Chapters I and II have two different Hilbert space basis,

$$e^{2\pi i\nu x}\, e^{2\pi i(nax+by)}\, e^{2\pi inz} \quad \text{and} \quad \psi_{n\nu}\, e^{2\pi i(anx+by)},$$

the first involving discontinuous functions, the second, involving C^∞ functions and Jacobi theta functions. We will begin this chapter by developing some formulas that show the relation between these two bases. One might guess that $\psi_{n\nu} = e^{2\pi i\nu x}\,\psi_{n0}$, but this is not quite right. Our computation will also provide us with another explicit form for the action of $R(N)$ on $H_{n\nu}$. Since

$$R(\tfrac{b}{2},\tfrac{a}{2},c)(\psi_{n\nu}\cdot g) = R(\tfrac{b}{2},\tfrac{a}{2},c)(\psi_{n\nu})\cdot R(\tfrac{b}{2},\tfrac{a}{2},c)(g),$$

$g \in L^2(\mathbf{T}^2,\tfrac{1}{n})$ or $g \in C^k(\mathbf{T}^2,\tfrac{1}{n})$ and $R(\tfrac{b}{2},\tfrac{a}{2},c)g(x,y) = g(x+\tfrac{b}{2},y+\tfrac{a}{2})$, the essential part of the second problem is the computation of $R(\tfrac{b}{2},\tfrac{a}{2},c)\psi_{n\nu}$. The original problem of comparing bases will fall out of this computation.

Now the computation of $R(\tfrac{b}{2},\tfrac{a}{2},c)\psi_{n\nu}$, because we have explicit intertwining operators between H_{n0} and $H_{n\nu}$ that carry ψ_{n0} to $\psi_{n\nu}$, reduces to the special case $\nu = 0$. Recall that

$$\psi_{n0}(x,y,z) = \sum_{n\in Z} \exp n\pi i(im^2+2m\zeta)\; e^{2\pi inz}\, e^{-\pi ny^2}$$

and $R(\tfrac{b}{2},\tfrac{a}{2},0)\psi_{n0}$ may be written after some elementary operations as

$$R(\tfrac{b}{2},\tfrac{a}{2},0)\psi_{n0}(x,y,z) = e^{2\pi inz}\, e^{-\pi ny^2}\, e^{2\pi iny\frac{b}{2}}\, e^{-2\pi ny\frac{a}{2}} \sum_{m\in Z} \exp n\pi i(i(m+\tfrac{a}{2})^2 + 2m(\zeta+\tfrac{b}{2}))$$

which after further algebraic operations becomes

1. $$R(\tfrac{b}{2},\tfrac{a}{2},0)\psi_{n,0}(x,y,z) = e^{2\pi inz}\, e^{-\pi ny^2}\, e^{2\pi in(y\frac{b}{2}-\frac{a}{2}x)}\, C\cdot$$

$$\sum_{m\in Z} \exp n\pi i(i(m+\tfrac{a}{2})^2 + 2(m+\tfrac{a}{2})(\zeta+\tfrac{b}{2}))$$

where $C = e^{2\pi ni\frac{ab}{4}}$. Thus $R(\tfrac{b}{2},\tfrac{a}{2},0)\psi_{n,0}(x,y,z) = CM_n[{a \atop b}](i)(\theta_i[{a \atop b}](n\zeta,ni))$.

But $M_n\left[\begin{smallmatrix}0\\0\end{smallmatrix}\right](i)(\theta\left[\begin{smallmatrix}\frac{2\nu}{n}\\0\end{smallmatrix}\right](n\zeta,ni)) = \psi_{n\nu}$. Hence

2. $\psi_{n\nu} = e^{2\pi i\nu x}\cdot R(0,\frac{\nu}{n},0)(\psi_{n,0})$.

Formulas 1 and 2 combine to settle the two questions we raised at the beginning of this chapter. For the rest of this chapter we will confine our attention to H_1 and relate the results of functions in H_1 to classical results about Jacobi theta functions. We will begin by giving applications of formula (1) in the special case where $n = 1$. To simplify notation for the rest of this chapter we will let $\psi_{10} = \psi$.

We will begin by deriving the classical reduction formula for first order Jacobi theta functions.

$$\theta\begin{bmatrix}2\nu+a\\2\nu'+b\end{bmatrix}(\zeta,i) = (-1)^{a\nu'}\ \theta\begin{bmatrix}a\\b\end{bmatrix}(\zeta,i)$$

where $a,b,\nu,\nu' \in \mathbf{Z}$. We will begin by assuming at first only that $\nu,\nu' \in \mathbf{Z}$. The proof of the reduction formula depends on formula (1). First,

3. $R(\frac{b}{2}+\nu',\frac{a}{2}+\nu,\frac{1}{4}(2\nu+a)(2\nu'+b))\psi(x,y,z)$

$$= \psi(\nu+\frac{b}{2}+\nu',y+\frac{a}{2}+\nu,z+\nu\nu'+\tfrac{1}{2}(a\nu'+b\nu)+\frac{ab}{4}+y(\frac{b}{2}+\nu')),$$

and

4. $R(\frac{b}{2},\frac{a}{2},\frac{ab}{4})\ \psi(x,y,z) = \psi(x+\frac{b}{2},y+\frac{a}{2},\frac{ab}{4}+z+y\,\frac{b}{2})$.

Since ν' and ν are integers $\gamma = (-\nu',-\nu,0) \in \Gamma$ and so we have the right side of 3) equals

$$\psi(x+\frac{b}{2},y+\frac{a}{2},z+\nu\nu'+\tfrac{1}{2}(a\nu'+b\nu)+\frac{ab}{4}+y(\frac{b}{2}+\nu')-\nu(x+\frac{b}{2}+\nu'))$$

$$= \psi(x+\frac{b}{2},y+\frac{a}{2},\frac{ab}{4}+z+y\,\frac{b}{2})\ e^{\pi i a\nu'}\ e^{2\pi i\nu' y}\ e^{-2\pi i\nu x}.$$

This shows that for $\nu,\nu' \in \mathbf{Z}$, we have

$$\theta\left[\begin{smallmatrix}a\\b\end{smallmatrix}\right](\zeta,i) = e^{-\pi i a\nu'}\ \theta\left[\begin{smallmatrix}2\nu+a\\2\nu'+b\end{smallmatrix}\right](\zeta,i)$$

For $a \in \mathbf{Z}$ this reduces to the reduction formula.

We will now define the reduced characters where a and b are 0 or 1 and denote these by $\left[\begin{smallmatrix}\epsilon\\\epsilon'\end{smallmatrix}\right]$. We will now prove the following important result.

$$\theta[{}^{\epsilon}_{\epsilon'}](-\zeta,i) = (-1)^{\epsilon\epsilon'}\,\theta[{}^{\epsilon}_{\epsilon'}](\zeta,i)\ .$$

We begin by noting that the mapping

$$A: (x,y,z) \to (-x,-y,z)$$

is an automorphism of N_3 that carries Γ onto itself. It thus induces a mapping A^* of $\Gamma\backslash N_3$ onto itself and it is readily checked that

$$A^*\psi(x,y,z) = \psi(-x,-y,z) = \psi(x,y,z)\ .$$

We will now apply formula (1). First

$$\theta[{}^{\epsilon}_{\epsilon'}](\zeta,i) = e^{-2\pi iz}\,e^{-\pi y^2}\,e^{2\pi i(\frac{\epsilon'}{2}y-\frac{\epsilon}{2}x)}\,R(\tfrac{\epsilon'}{2},\tfrac{\epsilon}{2},\tfrac{\epsilon\epsilon'}{4})\psi(x,y,z)$$

$$\theta[{}^{\epsilon}_{\epsilon'}](-\zeta,i) = e^{-2\pi iz}\,e^{-\pi y^2}\,e^{-2\pi i(\frac{\epsilon'}{2}y-\frac{\epsilon}{2}x)}\,R(\tfrac{\epsilon'}{2},\tfrac{\epsilon}{2},\tfrac{\epsilon\epsilon'}{4})\psi(-x,-y,z)\ .$$

Now

$$R(\tfrac{\epsilon'}{2},\tfrac{\epsilon}{2},\tfrac{\epsilon\epsilon'}{4})\psi(-x,-y,z) = \psi(-x+\tfrac{\epsilon'}{2},\ -y+\tfrac{\epsilon}{2},\ z+\tfrac{\epsilon\epsilon'}{4}-y\tfrac{\epsilon'}{2})\ .$$

Let $\gamma = (-\epsilon',-\epsilon,0) \in \Gamma$ and use the invariance of ψ under left multiplication by elements of Γ and the fact that $\psi(-x,-y,z) = \psi(x,y,z)$ to conclude that

$$R(\tfrac{\epsilon'}{2},\tfrac{\epsilon}{2},\tfrac{\epsilon\epsilon'}{4})\psi(-x,-y,x) = \psi(x+\tfrac{\epsilon'}{2},\ y+\tfrac{\epsilon'}{2},\ z+\tfrac{\epsilon\epsilon'}{4}-\tfrac{y\epsilon'}{2}+\epsilon x-\tfrac{\epsilon'\epsilon}{2})\ .$$

This implies that

$$R(\tfrac{\epsilon'}{2},\tfrac{\epsilon}{2},\tfrac{\epsilon\epsilon'}{4})\psi(x,y,z) = e^{2\pi i(\frac{\epsilon'\epsilon}{2}-\epsilon x+\epsilon' y)}\,R(\tfrac{\epsilon'}{2},\tfrac{\epsilon}{2},\tfrac{\epsilon\epsilon'}{4})\psi(-x,-y,z)\ .$$

Using formula 1 we have

$$\theta[{}^{\epsilon}_{\epsilon'}](-\zeta,i) = (-1)^{\epsilon\epsilon'}\,\theta[{}^{\epsilon}_{\epsilon'}](\zeta,i)\ .$$

We will now use the invariance of $R(a,b,c)\psi(x,y,z)$ under left multiplication by elements of Γ to prove the classical theta function identities with theta characteristics.

A) $\qquad \theta[{}^{\epsilon}_{\epsilon'}](\zeta+1,i) = (-1)^{\epsilon}\,\theta[{}^{\epsilon}_{\epsilon'}](\zeta,i)$,

B) $\qquad \theta[{}^{\epsilon}_{\epsilon'}](\zeta+i,i) = (-1)^{\epsilon'}\,e^{\pi i(-2\zeta-i)}\,\theta[{}^{\epsilon}_{\epsilon'}](\zeta,i)$.

Proof of A). $R(\frac{\epsilon'}{2}, \frac{\epsilon}{2}, \frac{\epsilon\epsilon'}{4})\psi(x,y,z) = \psi(x + \frac{\epsilon'}{2}, y + \frac{\epsilon}{2}, z + \frac{\epsilon\epsilon'}{4} + y\frac{\epsilon'}{2})$ is invariant under left multiplication by $(1,0,0) \in \Gamma$. Hence

$$R(\frac{\epsilon'}{2}, \frac{\epsilon}{2}, \frac{\epsilon\epsilon'}{4})\psi(x,y,z) = \psi(x + 1 + \frac{\epsilon'}{2}, y + \frac{\epsilon}{2}, z + \frac{\epsilon\epsilon'}{4} + y\frac{\epsilon'}{2}) .$$

By formula 1 :

$$\theta[{}^{\epsilon}_{\epsilon'}](\zeta+1,i) = e^{-2\pi i z} e^{2\pi i(\frac{\epsilon'}{2}y - \frac{\epsilon}{2}(x+1))} \psi(x + 1 + \frac{\epsilon'}{2}, y + \frac{\epsilon}{2}, z + \frac{\epsilon\epsilon'}{4} + y\frac{\epsilon'}{2})$$

$$= e^{\pi i \epsilon} \theta[{}^{\epsilon}_{\epsilon'}](\zeta,i) = (-1)^{\epsilon} \theta[{}^{\epsilon}_{\epsilon'}](\zeta,i) .$$

Proof of B). We will derive formula B by translating $\psi(x + \frac{\epsilon'}{2}, y + \frac{\epsilon}{2}, z + \frac{\epsilon\epsilon'}{4} + y\frac{\epsilon'}{2})$ on the left by $(0,1,0)$. The result of this operation is

$$\psi(x + \frac{\epsilon'}{2}, y + 1 + \frac{\epsilon}{2}, z + \frac{\epsilon\epsilon'}{4} + y\frac{\epsilon'}{2} + x + \frac{\epsilon'}{2}) .$$

Hence

$$\theta[{}^{\epsilon}_{\epsilon'}](\zeta+i,i) = e^{-2\pi i z} e^{2\pi i(\frac{\epsilon'}{2}(y+1) - \frac{\epsilon}{2}x)} e^{-\pi(y+1)^2}$$

$$\cdot \psi(x + \frac{\epsilon'}{2}, y + 1 + \frac{\epsilon}{2}, z + \frac{\epsilon\epsilon'}{4} + (y+1)\frac{\epsilon'}{2} + x + \frac{\epsilon'}{2}) .$$

Now as we discussed earlies in this chapter we may look at H_1 as $\mathcal{J}\Theta(L^2(\mathbb{R}))$ or as $\psi L^2(\mathbb{T}^2,1)$. A. Weil in [7] used the unitary representation theory of N to produce unitary representations of certain groups related to $SL(2,\mathbb{R})$. Some of Weil's ideas can be given another formulation that shows their deep connection with the classical theory of transformation theory of Jacobi theta functions by using the space H_1. We will spend the next part of this chapter presenting the details of this assertion.

Let us begin by recalling the action of $SL(2,\mathbb{R})$ in N and by reproducing a computation given by Weil. As in Chapter I, let $L(N)$ be the Lie algebra of N with basis X,Y,Z. Then the linear transformation

$$\begin{aligned} Y &\to aY + cX \\ X &\to bY + dX \\ Z &\to Z \end{aligned} \qquad \begin{pmatrix} a & b \\ c & d \end{pmatrix} \in SL_2(Z)$$

is an automorphism of the algebra structure on $L(N)$ and, since N is simply connected gives rise to an automorphism σ of N which may be written explicitly as

$$\sigma(x,y) = (cy + dx,\ bx + ay,\ z + \tfrac{1}{2}(bdx^2 + 2bcxy + acy^2)) .$$

For latter convenience we will let $q_\sigma(x,y,z) = \frac{1}{2}(bdx^2 + 2bcxy + acy^2)$. Further note that σ acts trivially on the center of N . We note that the subgroup L of $SL_2(\mathbf{Z})$ that takes Γ onto itself consists of all σ satisfying the congruence relation $ac \equiv bd \equiv 0 \mod 2$. A glance at $q_\sigma(x,y,z)$ justifies this assertion.

Now arguing exactly as in Chapter I, if $\sigma \in L$ and $F \in L^2(\Gamma\backslash N)$ clearly $T_\sigma(F) = F \circ \sigma \in L^2(\Gamma\backslash N)$ and the mapping $T_\sigma\colon L^2(\Gamma\backslash N) \to L^2(\Gamma\backslash N)$ is a unitary operator. Further, since σ acts trivially on the center of N, it follows that $T_\sigma(H_n) = H_n$ for all $n \in \mathbf{Z}$. We will now see one way of describing $T_\sigma(H_1)$.

<u>Theorem III.1</u>: Let $F \in H_1$; $F = \mathscr{J}\Theta f$, $f \in L^2(\mathbf{R})$. For

$$\sigma = \begin{pmatrix} a & b \\ c & d \end{pmatrix} \in \Gamma .$$

$$T_\sigma(F) = e^{2\pi iz} \sum_{m\in\mathbf{Z}} f(m+ay+bx)\, e^{2\pi i(q_\sigma(x,y)+m(cy+dx))}$$

and

$$T_\sigma(F) = e^{2\pi iz} \sum_{m\in\mathbf{Z}} \beta(m+y)\, e^{2\pi imx} , \qquad \beta \in L^2(\mathbf{R})$$

where β is given by

$$\beta(t) = \sum_{m\in\mathbf{Z}} \int_0^1 f(m+at+bx)\, e^{2\pi i(q_\sigma(x,t)+m(dx+ct))}\, dx .$$

The proof of Theorem III.1 requires only a technical modification of the nonstandard proof of Plancherel's theorem as given in Chapter I and so will be omitted.

Our desire is now to specify the action of L on H_1 using the description of H_1 in terms of Jacobi theta functions. Clearly if $\psi g \in H_1$, then $T_\sigma(\psi g) = T_\sigma(\psi)\cdot T_\sigma(g)$. Since $g \in L^2(\mathbf{T}^2,1)$, $T_\sigma(g)$ is classically known and so the only point of the discussion is to compute $T_\sigma(\psi)$. Until now the theta period τ has played very little role in our discussion. It will now play a crucial role. Let us pause to recall some notation and formulas from the previous chapters.

We defined, for $a,b \in \mathbb{R}$ and $\mathrm{Im}(\tau) > 0$,

$$\theta[{a \atop b}](\zeta,\tau) = \sum_{m\in\mathbf{Z}} \exp \pi i(\tau(m+\tfrac{a}{2})^2 + 2(m+\tfrac{a}{2})(\zeta+b))$$

and noted that

$$M_1[{a \atop b}](\tau)(\theta[{a \atop b}](\zeta,\tau)) \in H_1 \quad .$$

We will now introduce the notation $\theta_3(\zeta,\tau) = \theta[{0 \atop 0}](\zeta,\tau)$ and

$$\psi_3(x,y,z|\tau) = M_1[{0 \atop 0}](\tau)(\theta_3(\zeta,\tau)) = e^{2\pi i z} \sum_{m\in\mathbf{Z}} e^{\pi i\tau(y+m)^2} e^{2\pi i m x} \quad .$$

In this notation our problem reduces to computing $T_\sigma(\psi_3(x,y,z|\tau)$. Now the group L acts on the complex upper half plane by the fractional linear transformation

$$\sigma(\tau) = \frac{a\tau + c}{b\tau + d} \qquad \sigma = \begin{pmatrix} a & b \\ c & d \end{pmatrix} \in \Gamma \quad \mathrm{Im}(\tau) > 0 \ .$$

[Note: we are really letting the transpose of σ act in the usual notation]. For $\sigma \in \Gamma$ we can consider the two functions

$$\psi_3(\sigma(x,y,z|\tau) \quad \text{and} \quad \psi_3(x,y,z|\sigma(\tau)) \quad .$$

The transformation theory of Jacobi theta functions will turn out to be the study of the relationship between these two functions. This study will rest on the group generator structure of $SL_2(\mathbf{Z})$.

Lemma III.2: For $\sigma = \begin{pmatrix} 1 & 0 \\ 2 & 1 \end{pmatrix}$, we have

$$\psi_3(\sigma(x,y,z)|\tau) = \psi_3((x,y,z)|\sigma(\tau)) \ .$$

Proof: $\psi_3(\sigma(x,y,z)|\tau) = \psi_3((x+2y,y,z+y^2)|\tau) = e^{2\pi i(z+y^2)} \sum_{m\in\mathbf{Z}} e^{\pi i\tau(y+m)^2} e^{2\pi i m(x+2y)}$

$$= e^{2\pi i z} \sum_{m\in\mathbf{Z}} e^{\pi i(\tau+2)(y+m)^2} e^{2\pi i m x}$$

$$= \psi_3(x,y,z|\sigma(\tau)) \ .$$

Lemma III.3: Let $\sigma = \begin{pmatrix} 0 & 1 \\ -1 & 0 \end{pmatrix}$ then

$$\psi_3(\sigma(x,y,z|\tau) = \sqrt{\frac{i}{\tau}}\,\psi_3(x,y,z|\sigma(\tau)) \ .$$

The proof of this is essentially our proof of the Plancherel theorem combined with the classical result that

$$\int_{-\infty}^{\infty} e^{-ax^2+2xy}\,dx = \sqrt{\frac{\pi}{a}}\; e^{\frac{y^2}{a}} \quad , \ re(a) > 0$$

and so we will omit these calculations.

Corollary III.4: $\theta_3(\zeta,\tau+2) = \theta_3(\zeta,\tau)$.

Corollary III.5: $\theta_3(\zeta,\tau) = \sqrt{\frac{i}{\tau}}\; e^{-\pi i \frac{\zeta^2}{\tau}}\,\theta_3(\frac{\zeta}{\tau}\ ,\ -\frac{1}{\tau})$.

We will present the proof of the second corollary only, as that of the first follows in a similar manner.

Proof of Corollary III.5: Since

$$\psi_3(x,y,z) = e^{2\pi i z}\; e^{\pi i \tau y^2}\;\theta_3(x+\tau y|\tau) \ .$$

Lemma III.3 implies on setting $y = z = 0$ that

$$e^{\pi i \tau x^2}\;\theta_3(\tau x|\tau) = \sqrt{\frac{i}{\tau}}\;\theta_3(x|-\frac{1}{\tau}) \ .$$

Setting $\alpha = \tau x$, we have

$$e^{\pi i \alpha^2/\tau}\;\theta_3(\alpha|\tau) = \sqrt{\frac{i}{\tau}}\;\theta_3(\frac{\alpha}{\tau}|-\frac{1}{\tau})$$

holds for all $\alpha = \tau x$, $x \in \mathbb{R}$. Since both sides are entire functions of α , the equality must hold for all values of α .

We will now assume facts about the generalized Legendre symbol $(\frac{P}{Q})$ and the Gauss sum $G(a,b)$ as presented, for instance, in S. Lang [5] pages 76 to 90.

Theorem III.6: Let $\sigma = \begin{pmatrix} a & b \\ c & d \end{pmatrix} \in L$. Then

$$\psi_3(\sigma(x,y,z)|\tau) = k(\sigma|\tau)\;\psi_3(x,y,z|\sigma(\tau))$$

where

$$k(\sigma|\tau) = \begin{cases} (\frac{b}{d})\ i^{-d/2}\sqrt{i/(\tau b+d)} & d \text{ odd} \\ (\frac{d}{b})\ i^{(b-1)/2}\sqrt{i/(\tau b+d)} & d \text{ even} \end{cases}$$

Proof: The case $b = 0$ follows directly by repeated applications of Lemma III.2. Assume $b > 0$. Then by Theorem III.1

$$\psi_3(\sigma(x,y,z)|\tau) = e^{2\pi iz} \sum_{m\in Z} g(y+m)\ e^{2\pi imx}$$

where

$$g(y) = \int_0^1 e^{2\pi iq_\sigma(x,y)} \sum e^{\pi i\tau(ay+bx+m)^2 + 2\pi im(cy+dx)}\, dx$$

where the convergence is uniform convergence. Hence

$$g(y) = e^{\pi i \frac{a}{b} y^2} \sum_{n \bmod b} e^{-\pi i \frac{d}{b} n^2} \sum_{m=n \bmod b} \frac{1}{b} \int_{ay+m}^{ay+m+b} e^{\pi i(\tau+\frac{d}{b})x^2 - 2\pi i\frac{y}{b} x}\, dx$$

$$= e^{\pi i \frac{a}{b} y^2} \left(\frac{1}{b} \sum_{n \bmod b} e^{-\pi i \frac{d}{b} n^2}\right) \int_{-\infty}^{\infty} e^{\pi i(\tau+\frac{d}{b})x^2 - 2\pi i\frac{y}{b} x}\, dx .$$

$$\text{If}\quad k(\sigma|\tau) = \begin{cases} \frac{1}{2b}\, G(-d,2b)\sqrt{\frac{bi}{\tau b+d}} & d \text{ odd} \\ \frac{1}{b}\, G(-\frac{d}{2},b)\sqrt{\frac{bi}{\tau b+d}} & b \text{ odd} \end{cases}$$

we have

$$g(y) = k(\sigma|\tau)\ e^{\pi i\sigma(\tau)y^2} .$$

It then follows from the definition of $\psi_3(x,y,z|\sigma(\tau))$ that

$$\psi_3(\sigma(x,y,z)|\tau) = k(\sigma,\tau)\ \psi_3(x,y,z|\sigma(\tau)) .$$

The theorem for $b > 0$ then follows by applying the basic facts about Gauss sums.

To consider the case $\sigma = \begin{pmatrix} a & b \\ c & d \end{pmatrix} \in L$, $b < 0$, write

$$\sigma = \begin{pmatrix} -a & -b \\ -c & -d \end{pmatrix}\begin{pmatrix} -1 & 0 \\ 0 & -1 \end{pmatrix}$$

and notice that $k(-\sigma|\tau) = k(\sigma|\tau)$ and

$$\psi_3(-I(x,y,z)|\tau) = \psi_3(x,y,z|\tau)$$

where $-I = \begin{pmatrix} -1 & 0 \\ 0 & -1 \end{pmatrix}$. This completes the proof of Theorem III.6.

Using the description of $g(y)$ in Theorem III.6, we may argue exactly as in the proof of Corollary III.5 to prove

Corollary III.7: Let $\sigma = \begin{pmatrix} a & b \\ c & d \end{pmatrix} \in L$. Then

$$\theta_3\left(\frac{\zeta}{bz+d}\Big|\sigma(\tau)\right) = k(\sigma|\tau)^{-1} e^{\pi i b\zeta^2/(b\tau+d)} \theta_3(\zeta|\tau) .$$

To finish our discussion we will discuss the inter-relation of L and $R(N)$ acting on H_1. Since

$$R(\tfrac{b}{2}, \tfrac{a}{2}, \tfrac{ab}{4})\psi(x,y,z|i) = M_1[{}^a_b](i)(\theta[{}^a_b](\zeta,i))$$

it is reasonable to expect that

$$R(\tfrac{b}{2}, \tfrac{a}{2}, \tfrac{ab}{4})\psi(x,y,z|\tau) = M_1[{}^a_b](\tau)(\theta[{}^a_b](\zeta,\tau))$$

which is true as can be verified by a direct computation. If we introduce the notation

$$\psi[{}^a_b](x,y,z|\tau) = R(\tfrac{b}{2}, \tfrac{a}{2}, \tfrac{ab}{4})\psi(x,y,z|\tau)$$

we have

$$\psi[{}^a_b](x,y,z|\tau) = e^{2\pi i(z+\frac{b}{2}y - \frac{a}{2}x)} e^{\pi i \tau t^2} \theta[{}^a_b](y\tau + x|\tau) .$$

Almost all the rest of this section now become trivial exercises once we note that $T_\sigma(R(\frac{b}{2}, \frac{a}{2}, \frac{ab}{4})\psi(x,y,z|\tau) = R(\sigma(\frac{b}{2}, \frac{a}{2}, \frac{ab}{4})\psi(\sigma(x,y,z|\tau))$ and the following easily verified reduction formula: for $a,b,u,v \in Z$

$$\psi\begin{bmatrix} 2u+a \\ 2v+b \end{bmatrix}(x,y,z|\tau) = e^{2\pi i(yv - ux)} \cdot (-1)^{va}\, \psi[{}^a_b](x,y,z|\tau) .$$

Accordingly we will state the results without proof.

Theorem III.8: For $\sigma = \begin{pmatrix} a & b \\ c & d \end{pmatrix} \in L$

$$\psi[\sigma[{}^{a}_{b}]](\sigma(x,y,z)|\tau) = K(\sigma,\tau)\, e^{2\pi i q_{\sigma}(a/2,b/2)}\, \psi[{}^{a}_{b}](x,y,z|\sigma(\tau)) .$$

This result may be rewritten in terms of the Jacobi theta functions as

Theorem III.9: For $\sigma = \begin{pmatrix} a & b \\ c & d \end{pmatrix} \in L$

$$\theta[\sigma[{}^{a}_{b}]](\zeta|\tau) = k(\sigma|\tau)\, e^{2\pi i q_{\sigma}(a/2,b/2)}\, e^{-\pi i b \frac{\zeta^2}{\tau b+d}} \cdot \theta[{}^{a}_{b}](\tfrac{\zeta}{\tau b+d}|\sigma(\tau)) .$$

Let us close this chapter by stating the result that enables us to define how $\sigma = \begin{pmatrix} 1 & 0 \\ 1 & 1 \end{pmatrix}$ acts on $\theta[{}^{a}_{b}](x,y,z|\tau)$ and so we have completed the action of $SL_2(\mathbf{Z})$ on the functions $\theta[{}^{a}_{b}](x,y,z|\tau)$.

Theorem III.10: $\psi[\sigma[{}^{a}_{b}]](\sigma(x,y,z)|\tau) = i^{-a}\, \psi[{}^{\;a}_{b+1}](x,y,z|\tau + 1)$.

This verification will be left to the reader.

APPENDIX TO CHAPTER III

COHOMOLOGY AND THETA FUNCTIONS

The transformation theory for $L \subset SL_2(\mathbb{Z})$ acting on $\psi(x,y,z|\tau)$ as discussed in Chapter II leads to certain objects from cohomology theory. We will devote this Appendix to a detailed statement of these results. We have placed this material in an appendix because the role of these objects in the general theory is not yet clear to us.

Let us begin by recalling the following basic definitions from the cohomology theory of groups. Let G be a group, A an abelian group, with group multiplication written multiplicatively, and $\mathrm{aut}(A)$ denote the group of automorphisms of A. We will fix a homomorphism $\rho: G \to \mathrm{aut}(A)$ and denote $\rho(g)(a)$, $g \in G$, $a \in A$, by $g \cdot a$. A map

$$\lambda : G \to A$$

is called a 1-cocycle if

$$\lambda(g_1 g_2) = \lambda(g_1)(g_1 \cdot \lambda(g_2)) \quad , \quad g_1, g_2 \in G .$$

We denote the collection of all such 1-cocycles by $Z^1(G,A,\rho)$ and note that it is a subgroup of the group of all maps of G into A.

A mapping $\lambda: G \to A$ is called a 1-coboundary if

$$\lambda(g) = g \cdot a(a)^{-1} \quad , \quad g \in G , \; a \in A$$

where a is fixed. It is easily seen that a 1-coboundary is a 1-cocycle and the subset of $Z^1(G,A,\rho)$ consisting of 1-coboundaries is a subgroup. The set of 1-coboundaries will be denoted by $B^1(G,A,\rho)$ and we define the 1-cohomology group by

$$H^1(G,A,\rho) = Z^1(G,A,\rho)/B^1(G,A,\rho) .$$

Let $\mathcal{H}$ denote the subset of $\mathbb{C}$ consisting of numbers with positive imaginary parts and as usual let $C(\mathcal{H})$ denote the continuous functions on $\mathcal{H}$. Clearly $C(\mathcal{H})$ is a ring with identity under pointwise addition and multiplication with the constant function 1 as the identity. The subset $C^{\times}(\mathcal{H})$ of

$\mathcal{C}(\mathcal{H})$ consisting of all functions whose range does not include 0 is the multiplicative group of units in $\mathcal{C}(\mathcal{H})$ and so $\mathcal{C}^{\times}(\mathcal{H})$ is an abelian group. The group L acts on $\mathcal{C}^{\times}(\mathcal{H})$ according to the rule

$$\sigma^{*}(a)(\tau) = a(\sigma(\tau)) \quad , \qquad \sigma \in L \, , \, a \in \mathcal{C}^{\times}(\mathcal{H}) \, .$$

The mapping $a \to \sigma^{*}(a)$, $a \in \mathcal{C}^{\times}(\mathcal{H})$ is a homomorphism ρ of L into $\mathrm{aut}(\mathcal{C}^{\times}(\mathcal{H}))$ and so we may form $B^{1}(L,\mathcal{C}^{\times}(\mathcal{H}),\rho)$.

<u>Theorem III.A1</u>: The mapping λ: $L \to \mathcal{C}^{\times}(\mathcal{H})$ defined by

$$\lambda(\sigma)(\tau) = K^{-1}(\sigma|\tau) \quad , \qquad \sigma \in L \, , \, \tau \in H$$

where

$$K(\sigma,\tau) = \frac{\psi(\sigma(x,y,z)|\tau)}{\psi(x,y,z|\sigma(\tau))}$$

is an element of $B^{1}(L,\mathcal{C}^{\times}(\mathcal{H}),\rho)$.

<u>Proof</u>: We must show that there exists a function $a \in \mathcal{C}^{\times}(\mathcal{H})$ such that $\lambda(\sigma) = \sigma^{*}(a)a^{-1}$ for all $\sigma \in L$. Define $a \in \mathcal{C}^{\times}(\mathcal{H})$ by

$$a(\tau) = \psi(0,0,0|\tau) \, .$$

Hence $\lambda(\sigma)(\tau) = \sigma^{*}(a)(\tau)(a(\tau))^{-1}$ and we have proven our assertion.

<u>Note</u>: $a(\tau)$ is usually called the theta series associated with the Jacobi theta function $\psi(a,y,z|\tau)$.

Let $B^{\times}(\pi^{2})$ denote the group of bounded measurable complex valued functions on π^{2} never taking the value 0 . The group L acts on $B^{\times}(\pi^{2})$ by the rule

$$\sigma(p)(x,y) = p(cy + dx, ay + bx)$$

where

$$\sigma = \begin{pmatrix} a & b \\ c & d \end{pmatrix} \in L \quad \text{and} \quad p \in B^{\times}(\pi^{2}) \quad .$$

Note that μ: $\sigma \to \sigma(p)$ is a homomorphism of L into $\mathrm{aut}(B^{\times}(\pi^{2}))$. This gives us the data necessary to define the cohomology group $H^{1}(L, B^{\times}(\pi^{2}), \mu)$.

<u>Theorem III.A2</u>: Fix $\tau \in \mathcal{H}$. Let $\sigma \in L$ and define λ_{τ} by

$$\lambda_\tau(\sigma)(x,y) = \frac{\psi(\sigma(x,y,z)|\tau)}{\psi(x,y,z|\tau)} .$$

Then λ_τ is a mapping of L into $B^\times(\pi^2)$ and is in $Z^1(L, B^\times(\pi^2),\mu)$. Further $\lambda_\tau = \lambda_i \bmod B^1(L, B^\times(\pi^2),\mu)$ and so λ_τ determines a unique 1-cohomology class.

Proof: It is a classical result (see [6]) that for each $\tau \in \mathfrak{H}$ the zeros of the function $\theta[{}^0_0](\zeta,\tau)$ occur in $\mathbb{C}$ at the points $\frac{1}{2} + \ell_1 + (\frac{1}{2} + \ell_2)\tau$, $\ell_1,\ell_2 \in \mathbb{Z}$ and furthermore the order of each zero of $\theta[{}^0_0](\zeta,\tau)$ is one.

The congruence conditions defining L immediately imply that $\theta[{}^0_0](cy + dx + (ay + bx)\tau|\tau)$ has the same zeros as $\theta[{}^0_0](\zeta,\tau)$. Hence

1) $$\lim \frac{\theta[{}^0_0](cy + dx + (ay + bx)\tau|\tau)}{\theta[{}^0_0](\zeta,\tau)}$$

is bounded.

2) $$\frac{\psi(\sigma(x,y,z)|\tau)}{\psi(x,y,z|\tau)}$$

is in $B^\times(\pi^2)$ . . This shows that

$$\lambda_\tau : L \to B^\times(\pi^2) .$$

[Care should be taken to note that even though 2) is a bounded function 1) is not bounded in $\mathbb{C}$, because $M_1[{}^0_0](\tau)$ is a function of y.]

From

$$\lambda_\tau(\sigma_1\sigma_2)(x,y) = \frac{\psi(\sigma_2\sigma_1(x,y,z)|\tau)}{\psi(x,y,z|\tau)}$$

we obtain

$$\lambda_\tau(\sigma_1\sigma_2) = \lambda_\tau(\sigma_1)[\sigma_1\cdot\lambda_\tau(\sigma_2)]$$

or

$$\lambda_\tau \in Z^1(L, B^\times(\pi^2),\mu) .$$

To prove λ_τ and λ_i are cohomologous, we first note that

$$a_\tau(x,y) = \frac{\psi(x,y,z|\tau)}{\psi(x,y,z|i)} \in B^\times(\pi^2) \quad .$$

From

$$\lambda_\tau(\sigma)(x,y) = \frac{\psi(\sigma(x,y,z)|\tau)}{\psi(x,y,z|\tau)}$$

we have

$$\lambda_\tau(\sigma)(x,y) = \frac{\psi(\sigma(x,y,z)|\tau)}{\psi(\sigma(x,y,z)|i)} \frac{\psi(x,y,z|i)}{\psi(x,y,z|\tau)} \lambda_i(\sigma)(x,y)$$

$$= a_\tau(\sigma(x,y))a_\tau(x,y)^{-1} \lambda_\tau(\sigma)(x,y) \ .$$

Hence

$$\lambda_\tau(\sigma) = (\sigma\cdot a_\tau)a_\tau^{-1} \lambda_i(\sigma) \ .$$

Since $(\sigma\cdot a_\tau)a_\tau^{-1} \in B^1(L, B^\times(\pi^2),\mu)$ we have proven our assertion.

CHAPTER IV

THETA FUNCTIONS AND DISTINGUISHED SUBSPACES

In Chapter III we concentrated our attention on the Hilbert space H_1 and first order theta functions. This was for two reasons. First, H_1 is the simplest meaningful example we could study and second, we will need the results giving the inter-relations between H_1 and first order theta functions in the study of the general case.

Chapter IV will concern itself with the space H_n, $n > 1$, and n^{th} order theta functions. Here, because H_n, $n > 1$, is not irreducible and the dimension of nth order theta functions, $\Theta_n[{a \atop b}](\tau,A)$, is n, the situation is more complicated than the case where $n = 1$. We have already, in Chapter I, used group theoretic techniques to prove the basic facts that H_n has multiplicity n and $\dim \Theta_n[{a \atop b}](\tau,A) = n$. Indeed, we used the representation theory of finite groups to prove that we could explicitly write

$$H_n = \oplus \sum_{\nu=0}^{n-1} H_{n\nu}$$

where the spaces $H_{n\nu}$ are irreducible under $R(N)$ and we gave a closed form to the orthogonal projection p_ν of H_n onto $H_{n\nu}$, $\nu = 0,\dots,n-1$. In this chapter we will begin by enlarging on the group theoretic results presented in Chapter I and then see how these group theoretic results can be related to theta functions.

From the infinite number of R-invariant irreducible subspaces of H_n, the first author and J. Brezin in [1] showed that the underlying group theory enables one to select a finite subcollection, called distinguished subspaces, that are more "natural" than the others. In [1] several characterizations of distinguished subspaces were presented. We will present here still another characterization, which although known to the authors of [1] was not presented in that paper because its advantages were not apparant at that time.

In Chapter I we defined an action of $SL_2(Z)$ as a group of automorphisms of N. For the subgroup $L \subset SL_2(Z)$, we saw that this action induced a unitary representation of L on $L^2(\Gamma\backslash N)$ and on each H_n. It is easy to see that for $0 \neq n \equiv 0 \bmod 2$ this representation extends to a unitary representation of all of $SL_2(Z)$ on H_n.

The group $GL_2(Q)$, Q the rationals, acts as a group of automorphisms of N by the following formula

$$(1) \qquad \sigma(x,y,z) = (cy + dx,\ ay + bx, \|\sigma\|z + q_\sigma(x,y))$$

where $\sigma = \begin{pmatrix} a & b \\ c & d \end{pmatrix} \in GL_2(Q)$, $\|\sigma\| = ad - bc$ and $q_\sigma(x,y) = \frac{1}{2}(bdx^2 + 2bcxy + acy^2)$.

We shall now define a sequence of subsets of $GL_2(Q)$ and in terms of their actions structure the theory of the H_n, $n > 1$. For $n \equiv 1 \bmod 2$ set B_n equal to the set of all integer matrices $\sigma = \begin{pmatrix} a & b \\ c & d \end{pmatrix}$ such that $\|\sigma\| = n$ and $ab \equiv cd \equiv 0 \bmod 2$. For $0 \neq n \equiv 0 \bmod 2$ set B_n equal to the set of all integer natrices σ such that $\|\sigma\| = n$. In the remainder of this paper we set $B = \bigcup_{n\neq 0} B_n$.

Each $\sigma \in B$ can be written $\sigma = \sigma_1 \cdot \sigma_2$ where $\sigma_1 \in SL_2(Z)$ and $\sigma_2(\Gamma)$ is a subgroup of finite index Γ. If $\|\sigma\| \equiv 1 \bmod 2$ then $\sigma_1 = \begin{pmatrix} 1 & 0 \\ 0 & 1 \end{pmatrix}$ and if $\|\sigma\| \equiv 0 \bmod 2$ then $\sigma_1 = \begin{pmatrix} 1 & 1 \\ 0 & 1 \end{pmatrix}$ or $\sigma_1 = \begin{pmatrix} 1 & 0 \\ 1 & 1 \end{pmatrix}$. Defining $\sigma(F)$, $F \in L^2(\Gamma\backslash N)$ by the formula

$$\sigma(F)(x,y,z) = F(\sigma(x,y,z))$$

it follows that σ_2 is an isometry of $L^2(\Gamma\backslash N)$ even when it is not a unitary operator. Hence if K is a Hilbert subspace of $L^2(\Gamma\backslash N)$, σ_2 defines a unitary operator of K and $\sigma_2(K)$ as abstract Hilbert spaces. It is easy to see that for $\sigma \in B_m$

$$\sigma(H_n) \subset H_{mn} \quad , \quad n,m \in Z - \{0\} .$$

<u>Theorem IV.1</u>: Let J be an R-invariant and irreducible subspace of $L^2(\Gamma\backslash N)$. Then $\sigma(J)$ is an R-invariant and irreducible subspace of $L^2(\Gamma\backslash N)$.

<u>Proof</u>: The Stone-Von Neumann theorem implies $J \subset H_n$ for some $n \in Z - \{0\}$. Taking $\sigma \in B_m$ it follows that $\sigma(J) \subset H_{nm}$ and the theorem easily follows from the above discussion and the readily verified formula below

$$R(a,b,c)\,\sigma(F)(x,y,z) = \sigma(R(\sigma^{-1}(a,b,c))(\sigma(F))(x,y,z)\ .$$

Now in Chapter I we proved that H_1 is an R-invariant irreducible subspace of $L^2(\Gamma\backslash N)$. Hence, by Theorem IV.1, the action of B on $L^2(\Gamma\backslash N)$ determines a collection of R-invariant and irreducible subspaces of $L^2(\Gamma\backslash N)$. Rather than beginning with the definition of a distinguished subspace of H_n, we will begin by introducing a slightly different collection of objects from which the distinguished subspaces may be derived. Set $\mathcal{D}_n^{\#}$ equal to the collection of R-invariant and irreducible subspaces of H_n of the form $\sigma(H_1)$, $\sigma \in B_n$. We call $\mathcal{D}_n^{\#}$ the set of super distinguished subspaces of H_n. The super distinguished subspaces, $\mathcal{D}_n^{\#}$, play a special role in the theory of distinguished subspaces of H_n. We shall show that associated to each $J \in \mathcal{D}_n^{\#}$ there corresponds an orthogonal direct sum decomposition of H_n into R-invariant irreducible subspaces consisting of distinguished subspaces of H_n and every distinguished subspace of H_n occurs in at least one such direct sum decomposition.

Let us recall some basic definitions from Chapter I. We will denote the left regular representation of N on $L^2(N)$ by L and recall that it is defined by the formula

$$L(g)F(h) = F(g^{-1}h) \quad , \quad g,h \in N \qquad F \in L^2(N).$$

For $n \neq 0$, $n \in Z$, set

$$\Lambda_n = \{g \in N \,|\, L(g)H_n = H_n\}\ .$$

A direct computation shows that Λ_n consists of the elements of N of the form $(\frac{a}{n}, \frac{b}{n}, z)$, $a,b \in Z$, $z \in \mathbb{R}$. Hence if Z denotes the center of N and ΓZ denoted the group generated by Γ and Z, then Λ_n contains ΓZ as a normal subgroup of index n^2. In fact, $\Lambda_n/\Gamma Z$ is isomorphic to $Z/nZ \oplus Z/nZ$. The representation L restricted to Λ_n defines the unitary representation of Λ_n in H_n discussed in detail in Chapter I. Recall from Chapter I that if η_n is the character of ΓZ given by

$$\eta_n(a,b,z) = e^{2\pi i n z} \quad , \quad a,b \in Z \ , \ z \in \mathbb{R}\ .$$

Then $L(g)R = \eta_n(g)F$, $g \in \Gamma Z$, $F \in H_n$.

Before we present the general case of the discussion that was used to define the subspaces $H_{n\nu}$ in Chapter I, we shall consider some of the algebraic theory of the groups

$$S_n = Z/nZ \oplus Z/nZ \quad , \quad n \geq 1 .$$

Theorem IV.2: Let G be a subgroup of S_n of order n. Then there exists integers $0 < a_1, b_1, b_2 < n$ such that $a_1 b_2 = n$ and G is the subgroup of S_n generated by the elements (b_1, a_1) and $(b_2, 0)$.

Proof: Let $p: S_n \rightarrow Z/nZ$ denote the projection map onto the second factor. Since every subgroup of a cyclic group is cyclic, $p(G)$ and $G \cap \ker p$ are cyclic groups. Write $p(G) = \hat{a}_1$ where $a_1 \in Z$ and $\hat{a}_1$ denotes the subgroup of Z/nZ generated by a_1 . Since $\hat{d} = \hat{a}_1$ in Z/nZ , where $d = G \subset D(a_1, n)$, the greatest common divisor of a_1 and n, it follows that we can choose $0 < a_1$ and a_1/n , where this notation means that a_1 is divisible by n. Then the order of $p(G)$ is a_1 . Choose $0 < b_1 \leq n$ so that $(b_1, a_1) \in G$. In the same way $\ker p \cap G = (b_2, 0)^{\wedge}$ where $0 < b_2$, b_2/n and the order of $\ker p \cap G = n/b_2$. Clearly the short exact sequence $0 \rightarrow \ker p \cap G \rightarrow G \rightarrow p(G) \rightarrow 0$ implies $n^2/a_1 b_2 = n$ or $a_1 b_2 = n$. Also G is generated by (b_1, a_1) and $(b_2, 0)$.

We shall denote by $\mathcal{E}_n$ the collection of all subgroups of S_n or order n.

We take the usual action of $SL_2(Z)$ on S_n defined by the formula

$$\sigma(u,v) = (au + bv, cu + dv), \ \sigma = \begin{pmatrix} a & b \\ c & d \end{pmatrix} \in SL_2(Z).$$

Let $n > 1$. For each integer $\ell > 1$ such that ℓ^2/n we define the subgroup $\Lambda(n,\ell)$ of S_n by

$$\Lambda(n,\ell) = (\ell,0)^{\wedge} \oplus (0,n/\ell)^{\wedge}.$$

Theorem IV.3: The groups $\Lambda(n,\ell)$ with ℓ^2/n form a complete set of representations for the space of orbits $\mathcal{E}_n/B_1$ whenever $n \equiv 1 \bmod 2$ and $\mathcal{E}_n/SL_2(Z)$ whenever $n \equiv 0 \bmod 2$.

Proof: To prove the uniqueness aspect of the theorem we suppose

$$\sigma((\ell,c)^{\wedge}\oplus(0,n/\ell))^{\wedge} = (\ell',0)^{\wedge}\oplus(0,n/\ell')^{\wedge}$$

for $\sigma = \begin{pmatrix} a & b \\ c & d \end{pmatrix} \in SL_2(Z)$. Then from

$$\sigma((\ell,0)) = (a\ell, c\ell) = (x_1\ell', y_1(n/\ell'))$$

for some $x_1, y_1 \in Z$, it follows that $n/a\ell - x_1\ell'$ and $n/c\ell - y_i(n/\ell')$. Since ℓ'/n, we get $\ell'/\ell a$ and $\ell'/\ell c$. Using $GCD(a,c) = 1$ this latter fact implies ℓ'/ℓ. In the same way ℓ/ℓ' and hence $\ell = \ell'$.

Take $G \in \mathcal{E}_n$ and write as above $G = (b_2,0)^{\wedge} + (b_1,a_1)^{\wedge}$. Let $\ell = GCD(a_1,b_1,b_2)$. A standard number theoretic fact implies

$$GCD\left(\frac{-b_1 + b_2x}{\ell}, \frac{a_1}{\ell}\right) = 1$$

for some $x \in Z$. Hence

$$\sigma = \begin{pmatrix} r & s \\ \dfrac{a_1}{\ell}, & \dfrac{-b_1 + b_2x}{\ell} \end{pmatrix} \in SL_2(Z)$$

for some $r,s \in Z$. A simple computation shows $\sigma(G) = (\ell,0)^{\wedge}\oplus(0,n/\ell)^{\wedge}$. This completes the proof of the theorem when $n \equiv 0 \bmod 2$. For $n \equiv 1 \bmod 2$, since $\begin{pmatrix} 1 & 1 \\ 0 & 1 \end{pmatrix}$ and $\begin{pmatrix} 1 & 0 \\ n & 1 \end{pmatrix}$ for $\sigma(G)$ and either $\begin{pmatrix} 1 & 1 \\ 0 & 1 \end{pmatrix}\sigma \in B_1$ or $\begin{pmatrix} 1 & 0 \\ n & 1 \end{pmatrix}\sigma \in B_1$, the theorem is verified in this case as well.

Using a proof similar to the proof of Theorem IV.2, we can prove the following.

Theorem IV.3: The elements $\begin{pmatrix} \ell\mu & 0 \\ 0 & \ell \end{pmatrix}$, $n = \ell^2\mu$ form a complete system of representatives for the double coset space $B_1\backslash B_n/B_1$ whenever $n \equiv 1 \bmod 2$ and $SL_2(Z)\backslash B_n/SL_2(Z)$ whenever $n \equiv 0 \bmod 2$.

We return to the central topic. Set $\mathcal{E}_n^{\#}$ equal to the subgroups $\Delta^{\#} \subseteq \Lambda_n/\Gamma Z$ of order $n \geq 1$. For each $\Delta^{\#} = \Delta/\Gamma Z \in \mathcal{E}_n^{\#}$ and $\sigma \in B_m$ we define $\sigma^{-1}(\Delta^{\#}) = \sigma^{-1}(\Delta)/\Gamma Z$.

Theorem IV.4: For each $\Delta^{\#} \in \mathcal{E}^{\#}_{n}$ and $\sigma \in B_m$, $\sigma^{-1}(\Delta^{\#}) \in \mathcal{E}^{\#}_{nm}$. Furthermore, $B_m^{-1}(\mathcal{E}^{\#}_n) = \mathcal{E}^{\#}_{nm}$ where $B_m^{-1}(\mathcal{E}^{\#}_n) = \{\sigma^{-1}(\Delta^{\#}): \sigma \in B_m, \Delta^{\#} \in \mathcal{E}^{\#}_n\}$.

Proof: Since $\sigma^{-1}(\Lambda_n) \subset \Lambda_{nm}$ and $\sigma^{-1}(\Gamma Z) \supset \Gamma Z$ it follows that $\sigma^{-1}(\Delta)$ is a subgroup of Λ_{nm} containing ΓZ. Also $\sigma^{-1}(\Delta^{\#}/\sigma^{-1}(\Gamma Z)^{\#}) = \Delta^{\#}$ which along with $\sigma^{-1}(\Gamma Z)/\Gamma Z = Z^2/\sigma(Z^2)$ implies $\sigma^{-1}(\Delta)/\Gamma Z \in \mathcal{E}^{\#}_{nm}$.

The proof is complete once we show that whenever $\Delta^{\#} = \Delta/\Gamma Z \in \mathcal{E}^{\#}_n$ there is a $\sigma \in B_n$ such that $\sigma^{-1}(\Gamma Z) = \Delta$. This follows immediately from Theorem IV.2.

As a particular case of Theorem IV.4 we have that the map $\sigma \to \sigma^{-1}(\Gamma Z)$ maps B_n onto $\mathcal{E}^{\#}_n$.

The next series of results relates the collection $\mathcal{E}^{\#}_n$ and the super distinguished subspaces $\mathcal{D}^{\#}_n$. This association is established by using the representation L to induce a family of representations $L(\Delta^{\#})$, $\Delta^{\#} \in \mathcal{E}^{\#}_n$, of $\Delta^{\#}$ on H_n. For simplicity we fix $n > 1$.

Take $\Delta^{\#} = \Delta/\Gamma Z \in \mathcal{E}^{\#}_n$. Then $\Delta = \Delta(a_1,b_1,b_2)\Gamma Z$ where $\Delta(a_1,b_1,b_2)$ is the group generated by the elements $(a_1/n, b_1/n, 0)$ and $(0,0,b_2/n)$ in Λ_n, for some integers $0 < a_1, b_1, b_2 \leq n$ and $a_1 b_2 = n$. Restricting L to $\Delta(a_1,b_1,b_2)$ defines a unitary representation of $\Delta(a_1,b_1,b_2)$ on H_n which vanishes on $\Gamma Z \cap \Delta(a_1,b_1,b_2)$. Hence L induces a unitary representation $L(\Delta^{\#})$ of $\Delta^{\#}$ on H_n.

For any R-invariant subspace J of H_n define the stabilizer Λ_J of J by

$$\Lambda_J = \{g \in \Lambda_n : L(g)J = J\} .$$

For any $\Delta^{\#} \in \mathcal{E}^{\#}_n$ define

$$J(\Delta^{\#}) = \{F \in H_n : L(\Delta^{\#})(g)F = F \quad \text{for all} \quad g \in \Delta^{\#}\} .$$

Theorem IV.5: Let $\sigma \in B_n$ and set $J = \sigma(H_1)$ and $\Delta^{\#} = \sigma^{-1}(\Gamma Z)/\Gamma Z$. Then

$$J = J(\Delta^{\#})$$

and (2)

$$\Delta^{\#} = \Lambda_J/\Gamma Z .$$

Proof: A simple calculation verifies the theorem.

The preceeding theorem established a bijection between $\mathcal{D}_n^{\#}$ and $\mathcal{E}_n^{\#}$. In one direction, associated to each super distinguished subspace $J \in \mathcal{D}_n^{\#}$ we have the stability group Λ_J of J (actually $\Lambda_J/\Gamma Z$); in the other direction to each $\Delta^{\#} \in \mathcal{E}_n^{\#}$ the bijection corresponds the eigenvalue one space of $L(\Delta^{\#})$ on H_n.

Next we associate to each $J \in \mathcal{D}_n^{\#}$ an orthogonal direct sum decomposition of H_n into R-invariant and irreducible subspaces. To do this we will establish for each $\Delta^{\#} = \Delta/\Gamma Z \in \mathcal{E}_n^{\#}$ a canonical correspondence between the characters of Δ extending, η_n, and the characters of $\Delta^{\#}$. Henceforth we will use $ch(\Delta^{\#})$ to denote the characters of $\Delta^{\#}$. Writing $\Delta = \Delta(a_1,b_1,b_2) \cdot \Gamma Z$ as above each character λ of Δ extending η_n restricts to a character of $\Delta(a_1,b_1,b_2)$ which vanishes on $\Gamma Z \cap \Delta(a_1,b_1,b_2)$ and so determines a character $\lambda^{\#}$ on $\Delta^{\#}$. In this way a bijection is established between the characters of Δ extending η_n and $ch(\Delta^{\#})$.

Let $\Delta^{\#} \in \mathcal{E}_n^{\#}$ and $J = J(\Delta^{\#})$. define

$$J(\Delta^{\#},\lambda^{\#}) = \{F \in H_n : L(\Delta^{\#})(g^{\#})F = \lambda^{\#}(g^{\#})F,\ g^{\#} \in \Delta^{\#}\},\ \lambda^{\#} \in ch(\Delta^{\#}) .$$

Theorem IV.6: For each $\lambda^{\#} \in ch(\Delta^{\#})$, $J(\Delta^{\#},\lambda^{\#})$ is an R-invariant and irreducible subspace of H_n and

$$(3) \qquad H_n = \oplus \sum_{\lambda^{\#} \in ch(\Delta^{\#})} J(\Delta^{\#},\lambda^{\#}) = \oplus \sum_{g \in \Lambda_n/\Delta^{\#}} L_g J .$$

In particular L_g, $g \in \Lambda_n$ is an intertwining operator for $R|J$ and $R|L_g J$.

Proof: Since $L(\Delta^{\#})$ is a unitary representation of the finite abelian group $\Delta^{\#}$ in H_n it follows from elementary representation theory that

$$H_n = \oplus \sum_{\lambda^{\#} \in ch(\Delta^{\#})} J(\Delta^{\#},\lambda^{\#})$$

is an orthogonal direct sum decomposition of H_n. Also it is easy to check that $J(\Delta^{\#},\lambda^{\#})$ is R-invariant. From the formula

$$L(g)R(h) = R(h)L(g), \quad g \in \Lambda_n \quad \text{and} \quad h \in N$$

it follows that $L(g)$ is an intertwining operator for $R|J$ and $R|L(g)J$. In particular since J is irreducible and R-invariant, $L(g)J$ is also. Hence we are done as soon as we know that the collection of spaces $J(\Delta^{\#},\lambda^{\#})$ is the same as the collection of spaces $L(g)J$. This is however, a trivial computation and in fact $L(g)J = J(\Delta^{\#},\lambda^{\#})$ where $\lambda^{\#}$ is induced by the character $h \to \eta_n[g,h]$ of Δ.

__Corollary:__ H_n is the orthogonal direct sum of n unitarily equivalent R-invariant and irreducible subspaces.

__Proof:__ Since the order of $\Lambda_n/\Delta^{\#}$ is n the result follows from (3).

__Corollary:__ If J is an irreducible R-invariant subspace of H_n such that $\Lambda_J/\Gamma Z$ has order n then there is a character $\lambda^{\#}$ on $\Lambda_J^{\#}$ such that $J = J(\Lambda_J^{\#},\lambda^{\#})$.

__Proof:__ Follows immediately by considering the unitary representation $L(\Lambda_J/\Gamma Z)$ on J.

The R-invariant and irreducible subspaces of H_n of the form $J = J(\Delta^{\#},\lambda^{\#})$ where $\Delta^{\#} \in \mathcal{E}_n^{\#}$ and $\lambda^{\#} \in ch(\Delta^{\#})$ will be called the distinguished subspaces of H_n, following [1], and the collection of all such distinguished subspaces of H_n will be denoted by $\mathcal{D}_n$.

The group B_1 acts as a group of permutations on $\mathcal{D}_n$, $n \equiv 1 \bmod 2$. For $\sigma \in B_1$, $J = J(\Delta^{\#},\lambda^{\#})$ this action is given by

$$\sigma(J) = J(\sigma^{-1}(\Delta^{\#}),\sigma(\lambda^{\#}))$$

where $\sigma(\lambda^{\#}) = \lambda^{\#} \circ \sigma$. There is an analogous statement for $n \equiv 0 \bmod 2$.

Now let us recall the following definitions and notations. Let $n \geq 1$ and $a,b, \in \mathbb{R}$. For each τ in the upper half complex plane, we defined an nth order theta function of period τ and characteristic $\begin{bmatrix} a \\ b \end{bmatrix}$ to be any continuous function φ on $\mathbb{C}$ satisfying the following functional equations.

$$\varphi(\zeta+1) = (-1)^a \varphi(\zeta) \quad , \quad \zeta \in \mathbb{C}$$

$$\varphi(\zeta+\tau) = (-1)^b e^{-\pi i n(2\zeta+\tau)}\varphi(\zeta) \quad , \quad \zeta \in \mathbb{C} .$$

Recall we denoted the space of all n^{th} order theta functions of period τ and characteristic $\left[\begin{smallmatrix}a\\b\end{smallmatrix}\right]$ by $\Theta_n\left[\begin{smallmatrix}a\\b\end{smallmatrix}\right](\tau)$ and the subspace of entire functions (argument ζ) by $\Theta_n\left[\begin{smallmatrix}a\\b\end{smallmatrix}\right](\tau,A)$.

In Chapter II it was proved that the map $M_n\left[\begin{smallmatrix}a\\b\end{smallmatrix}\right](\tau)$ of $\Theta_n\left[\begin{smallmatrix}a\\b\end{smallmatrix}\right](\tau)$ defined by the formula

$$(M_n\left[\begin{smallmatrix}a\\b\end{smallmatrix}\right](\tau)\varphi)(x,y,z) = e^{2\pi i(\frac{b}{2}y - \frac{a}{2}x + nz)} e^{\pi i\tau ny^2} \varphi(x+\tau y), \quad (x,y,z) \in N$$

is a $\mathbb{C}$-vector space isomorphism of $\Theta_n\left[\begin{smallmatrix}a\\b\end{smallmatrix}\right](\tau)$ onto C_n, the continuous functions in H_n.

Set $A_n\left[\begin{smallmatrix}a\\b\end{smallmatrix}\right](\tau) = M_n\left[\begin{smallmatrix}a\\b\end{smallmatrix}\right](\tau)\,\Theta_n\left[\begin{smallmatrix}a\\b\end{smallmatrix}\right](\tau,A)$. For fixed τ and $\left[\begin{smallmatrix}a\\b\end{smallmatrix}\right]$ the classical result that $A_n\left[\begin{smallmatrix}a\\b\end{smallmatrix}\right](\tau)$ is an n-dimensional $\mathbb{C}$-vector space has been proved in Chapter II. Recall that $A_1\left[\begin{smallmatrix}a\\b\end{smallmatrix}\right](\tau)$ consists of all complex multiples of

$$\psi\left[\begin{smallmatrix}a\\b\end{smallmatrix}\right](x,y,z|\tau) = M_1\left[\begin{smallmatrix}a\\b\end{smallmatrix}\right](\tau)\,\theta\left[\begin{smallmatrix}a\\b\end{smallmatrix}\right](x,y,z|\tau)$$

where

$$\theta\left[\begin{smallmatrix}a\\b\end{smallmatrix}\right](\zeta|\tau) = \sum_{m\in Z} e^{\pi i\tau(m+\frac{a}{2})^2} e^{2\pi i(m+\frac{a}{2})(\zeta+\frac{b}{2})}, \quad \zeta = x + iy \in \mathbb{C}.$$

The remainder of this section is devoted to investigating the action of B_n on the spaces $A_n\left[\begin{smallmatrix}a\\b\end{smallmatrix}\right](\tau)$. For convenience we consider this solely for the spaces $A_n\left[\begin{smallmatrix}0\\0\end{smallmatrix}\right](\tau)$.

Let us begin by recalling the basic formulas

$$L(r,s,0)\,\psi\left[\begin{smallmatrix}a\\b\end{smallmatrix}\right](x,y,z|\tau) = e^{2\pi i(sx - ry)} e^{-2\pi i r(s+\frac{a}{2})} \cdot \psi\left[\begin{smallmatrix}a+2s\\b+2r\end{smallmatrix}\right](x,y,z|\tau). \tag{4}$$

$$R(r,s,0)\,\psi\left[\begin{smallmatrix}a\\b\end{smallmatrix}\right](x,y,z|\tau) = e^{-2\pi i r(a+2s)} \psi\left[\begin{smallmatrix}a+2s\\b+2r\end{smallmatrix}\right](x,y,z|\tau).$$

Let us now prove a generalization of a result in Chapter 3.

<u>Theorem IV.7</u>: For $\sigma = \left(\begin{smallmatrix}a & b\\c & d\end{smallmatrix}\right) \in B_n$,

$$\sigma(A_1\left[\begin{smallmatrix}0\\0\end{smallmatrix}\right](\tau)) = A_n\left[\begin{smallmatrix}0\\0\end{smallmatrix}\right](\sigma(\tau)) \cap \sigma(H_1).$$

In particular

$$\sigma(\psi[{}^0_0](x,y,z|\tau)) = M_n[{}^0_0](\sigma(\tau))\varphi(\zeta)$$

where $\varphi(\zeta)$ is the first order theta function of period $\sigma(\tau)$ and character $[{}^0_0]$ given by

$$\varphi(\zeta) = e^{\pi i b(b\tau+d)\zeta^2}\,\theta[{}^0_0]((b\tau+d)\zeta|\tau) .$$

Proof: It suffices to prove $\sigma(\psi[{}^0_0](x,y,z|\tau)) = M_n[{}^0_0](\sigma(\tau))\varphi(\zeta)$ since $\varphi(\zeta)$ as given clearly is in $\Theta_1[{}^0_0](\sigma(\tau),A)$.

By (1) we get

$$\sigma(\psi[{}^0_0](x,y,z|\tau)) = e^{2\pi inz}\cdot W(y,x)\theta[{}^0_0]((b\tau+d)\zeta|\tau)$$

where

$$W(y,x) = e^{\pi i(acy^2+2bcyx+bdx^2)}e^{\pi i\tau(ay+bx)^2}$$

and $\zeta = x + \sigma(\tau)y$. A simple computation shows

$$W(y,x) = e^{\pi i\sigma(\tau)ny^2}\,e^{\pi ib(b\tau+d)\zeta^2}$$

and the theorem has been proved.

The preceeding theorem described the action of B on the spaces

$$J_A[{}^0_0](\tau) = J \cap A_n[{}^0_0](\tau) , \quad J \in \mathcal{D}_n^\# .$$

Corollary: For $\sigma \in B_1$, $J \in \mathcal{D}_n^\#$,

$$\sigma(J_A[{}^0_0](\tau)) = \sigma(J)_A[{}^0_0](\sigma(\tau)) .$$

To extend the above remarks to arbitrary distinguished subspaces we must investigate the representation L acting on the collection $A_n[{}^0_0](\tau)$.

Theorem IV.8: Let Λ_n be the stabilizer subgroup of H_n , $n \geq 1$. Then

$$L(\gamma)A_n[{}^0_0](\tau) = A_n[{}^0_0](\tau) , \quad \gamma \in \Lambda_n .$$

Proof: If $F \in A_n[{}^0_0](\tau)$ then $F(x,y,z) = e^{2\pi inz}\,e^{\pi i\tau ny^2}\,\varphi(\zeta)$ where $\varphi(\zeta)$, $\zeta = x + \tau y$, is in $\Theta_n[{}^0_0](\tau,A)$.

Take $\gamma = (r/n,s/n,z) \in \Lambda_n$. Then

$$L(\gamma)F(x,y,z) = e^{2\pi in\zeta}\,e^{\pi i\tau s^2/n}\,e^{2\pi inx}\,e^{\pi i\tau ny^2}\,\psi(x+\tau y)$$

where

$$\psi(x+\tau y) = e^{2\pi i s(x+\tau y)} \varphi(x+\frac{r}{n} + \tau(y+\frac{s}{n})) .$$

Clearly $\psi(\zeta)$, $\zeta = x + \tau y$ is an nth order theta function of period τ and characteristic $[{0 \atop 0}]$. Also it is analytic. Hence we get the desired result.

We close by generalizing the result of Chapter I about the projections of $A_n[{0 \atop 0}](\tau)$ onto $H_{nr} \in \mathcal{D}_n$.

Theorem IV.9: For any $J \in \mathcal{D}_n$ denote by p_J the projection of H_n onto J . Then

$$p_J A_n[{0 \atop 0}](\tau) = J_A[{0 \atop 0}](\tau)$$

(5)

$$A_n[{0 \atop 0}](\tau) = \oplus \sum_{\gamma \in \Lambda_n/\Lambda_J} L(\gamma) J \quad A[{0 \atop 0}](\tau) .$$

Proof: For $F \in A_n[{0 \atop 0}](\tau)$ by Theorem 6 there is a $\gamma \in \Lambda_n$ such that $L(\gamma)J \in \mathcal{D}_n^{\#}$. Write $F = p_J(F) + G$ where G is in the orthogonal complement to J . Then $L(\gamma) F = L(\gamma)p_J(F) + L(\gamma)G$. By the previous theorem it follows that $L(\gamma)p_J(F) \in A_n[{0 \atop 0}](\tau)$ if and only if $p_J(F)$ is. Hence we may assume without any loss of generality that $J \in \mathcal{D}_n^{\#}$.

Let $\Delta^{\#} = \Lambda_J/\Gamma Z$ then it is easy to see that

$$p_J(F) = 1/n(\sum_{\gamma^{\#} \in \Delta^{\#}} (L(\gamma^{\#})(\Delta^{\#})F)$$

which is in $A_n[{0 \atop 0}](\tau)$ again using the previous theorem. The second statement of (5) is now obvious and the theorem is verified.

In this part we give another characterization of the distinguished subspaces in terms of theta functions.

Setting

$$A_n[{0 \atop 0}](\tau) = M_n[{0 \atop 0}](\tau) \; \Theta_n[{0 \atop 0}](\tau, A)$$

we know that $A_n[{0 \atop 0}](\tau)$ is an n-dimensional complex vector space whose projection into any distinguished subspace J of H_n equals

$$J_A[{0 \atop 0}](\tau) = A_n[{0 \atop 0}](\tau) \cap J$$

which is a one dimensional complex vector space.

Our aim is to show that $J \in \mathcal{D}_n$ if and only if $J_A\left[\begin{smallmatrix}0\\0\end{smallmatrix}\right](\tau)$ contains the product of n first order theta functions on the Heisenberg group whose characteristics are related by the group structure. This characterizes distinguished subspaces in terms of the classical Jacobi theta functions.

Let $\Delta^{\#} = \Delta/\Gamma Z \in \mathcal{E}_n^{\#}$. Set $J = J(\Delta^{\#})$, i.e., the space of all $F \in H_n$ satisfying

$$L(\gamma^{\#})(\Delta^{\#})F = F \quad , \quad \gamma^{\#} \in \Delta^{\#} .$$

Recall $\Delta = \Delta(a_1,b_1,b_2)\Gamma Z$ where $0 < a_1,b_1,b_2 \le n$ with $a_1b_2 = |n|$ and $\Delta(a_1,b_1,b_2)$ the group generated by $(a_1/n,b_1/n,0)$ and $(0,b_2/n,0)$. Clearly

$$L(\gamma^{\#})(\Delta^{\#})F = L(\ell_1 \frac{a_1}{n} , \ell_1 \frac{b_1}{n} + \ell_2 \frac{b_2}{n}, \ell_1^2 \frac{a_1b_1}{2n^2}) F$$

for all $F \in H_n$,

$$\gamma = (\ell_1 \frac{a_1}{n}, \ell_1 \frac{b_1}{n} + \ell_2 \frac{b_2}{n} , z) \in \Delta .$$

This representation of $\Delta^{\#}$ on H_n extends to a representation of $\Delta^{\#}$ on the space of functions defined on N using exactly the same formula. This action will now be used to construct the kind of functions in $J_A\left[\begin{smallmatrix}0\\0\end{smallmatrix}\right](\tau)$ which are desired.

Theorem IV.10: Let J be an R-invariant and irreducible subspace of H_n. Then $J \in \mathcal{D}_n^{\#}$ if and only if it contains a function of the form

$$(6) \qquad \prod_{\substack{0\le\ell_1<b_2\\0\le\ell_2<a_1}} \psi \begin{bmatrix} 2(\ell_1 \frac{b_1}{n} + \ell_2 \frac{b_2}{n}) \\ 2\,\ell_1 \frac{a_1}{n} \end{bmatrix} (x,y,z|\tau)$$

for some integers $0 < a_1,b_1,b_2 \le n$ with $a_1b_2 = n$. In case J contains a function of this form then the stability group of J is $\Delta^{\#} = \Delta(a_1,b_1,b_2)\Gamma Z/\Gamma Z$.

Proof: Suppose $J \in \mathcal{D}_n^{\#}$ and has stability group $\Delta^{\#} = \Delta(a_1,b_1,b_2)\Gamma Z/\Gamma Z$. Set

$$G = \prod_{\gamma^{\#}\in\Delta^{\#}} L(\gamma^{\#})(\Delta^{\#})\ \psi\begin{bmatrix}0\\0\end{bmatrix}(x,y,z|\tau) \quad .$$

Then

$$G = M(y,x) \prod_{\substack{0\le\ell_1<b_2\\0\le\ell_2<a_1}} \psi\begin{bmatrix}2(\ell_1 \frac{b_1}{n} + \ell_2 \frac{b_2}{n})\\ 2\ell_1 \frac{a_1}{n}\end{bmatrix}(x,y,z|\tau) \; .$$

where

$$M(y,x) = \prod_{\substack{0\le\ell_1<b_2\\0\le\ell_2<a_1}} e^{2\pi i\ell_1^2 \frac{a_1 b_1}{2n^2}} \prod_{\substack{0\le\ell_1<b_2\\0\le\ell_2<a_1}} e^{2\pi i\left[\left(\ell_1 \frac{b_1}{n} + \ell_2 \frac{b_2}{n}\right)x-\ell_1 \frac{a_1}{n} y\right]}$$

Clearly $L(\gamma^{\#})(\Delta^{\#})G = G$ for all $\gamma^{\#} \in \Delta^{\#}$. Hence since

$$M(y+ \frac{b_1}{n}, x+ \frac{a_1}{n}) = M(y,x) \cdot \prod_{\substack{0\le\ell_1<b_2\\0\le\ell_2<a_1}} e^{2\pi i\left[\left(\ell_1 \frac{b_1}{n} + \ell_2 \frac{b_2}{n}\right) \frac{a_1}{n} - \ell_1 \frac{a_1}{n} \frac{b_1}{n}\right]}$$

$$= M(y,x) \prod_{\substack{0\le\ell_1<b_2\\0\le\ell_2<a_1}} e^{2\pi i \frac{\ell_2}{n}}$$

$$= M(y,x)$$

and

$$M(y + \frac{b_2}{n}, x) = M(y,x) \prod_{\substack{0\le\ell_1<b_2\\0\le\ell_2<a_1}} e^{-2\pi i\ell_1 \frac{a_1 b_2}{n^2}}$$

$$= M(y,x)$$

it follows that

$$L(\gamma^{\#})(\Delta^{\#}) \frac{G(x,y,z|\tau)}{M(y,x)} = \frac{G(x,y,z|\tau)}{M(y,x)} \quad \text{for all} \quad \gamma^{\#} \in \Delta^{\#} \; .$$

Set

$$H = \prod_{\substack{0\le \ell_1<b_2\\ 0\le \ell_2<a_1}} \psi\begin{bmatrix} 2(\ell_1 \frac{b_1}{n} + \ell_2 \frac{b_2}{n}) \\ 2\,\ell_1 \frac{a_1}{n} \end{bmatrix}(x,y,z|\tau) .$$

Using the identities $\sum_{\substack{0\le \ell_1<b_2\\ 0\le \ell_2<a_1}} (\ell_1 \frac{b_1}{n} + \ell_2 \frac{b_2}{n}) = n - b_2 - b_1 + b_1 b_2$ and

$$\sum_{\substack{0\le \ell_1<b_2\\ 0\le \ell_2<a_1}} \ell_1 \frac{a_1}{n} = n - a_1$$

it is then easy to see that $H \in J_A\begin{bmatrix} n-b_2-b_1+b_1b_2 \\ n-a_1 \end{bmatrix}(\tau)$. Hence if $J \in \mathcal{D}_n^{\#}$ then J contains a function of the stated form. Conversely if J contains a function of this form then $J = J(\Delta^{\#})$, where $\Delta^{\#} = \Delta(a_1,b_1,b_2)\Gamma Z/\Gamma Z$.

<u>Corollary</u>: Let $J \in \mathcal{D}_n^{\#}$. Then the function

$$\prod_{\substack{0\le \ell_1<b_2\\ 0\le \ell_2<a_1}} \psi\begin{bmatrix} -1+ \frac{b_1+b_2-b_1b_2}{n} + 2(\ell_1 \frac{b_1}{n} + \ell_2 \frac{b_2}{n}) \\ -1+ \frac{a_1}{n} + 2\frac{\ell_1}{n} \end{bmatrix}(x,y,z|\tau) \in$$

$J_A\begin{bmatrix} 0 \\ 0 \end{bmatrix}(\tau)$ if and only if $\Delta^{\#} = \Delta(a_1,b_1,b_2)/\Gamma Z$, $0 < a_1,b_1,b_2 \le n$, $a_1 b_2 = n$, is the stability group of J .

For any integer $n \ge 1$ let H_n be the corresponding multiplicity space in $L^2(L\backslash N)$ and $A_n\begin{bmatrix} 0 \\ 0 \end{bmatrix}(\tau)$ the analytic subspace of H_n defined above.
For any super distinguished subspace $J \in \mathcal{D}_n^{\#}$ with stabilizer subgroup $\Delta^{\#}$ we have the orthogonal decomposition of H_n into R-invariant and irreducible subspaces given by

$$H_n = \sum_{\gamma \in \Lambda_n^{\#}/\Delta^{\#}} L(\gamma)J$$

and corresponding to this decomposition the decomposition of $A_n\begin{bmatrix} 0 \\ 0 \end{bmatrix}(\tau)$ into one dimensional complex vector spaces given by

$$A_n[{}^0_0](\tau) = \sum_{\gamma\in\Lambda_n^{\#}/\Delta^{\#}} L(\gamma)J\, A[{}^0_0](\tau)$$

where $L(\gamma)J\ {}_A[{}^0_0](\tau) = L(\gamma)J \cap A_n[{}^0_0](\tau)$.

Suppose $J = \sigma(H_1)$ where $\sigma \in B_n$. Then using Theorem 7 and Theorem 8 the underlying group structure determines a basis of $A_n[{}^0_0](\tau)$. Setting $\Psi = \sigma(\psi[{}^0_0](x,y,z|\tau'))$ where $\sigma(\tau') = \tau$, this natural basis is given by

$$L(\gamma)\ \Psi:\ \gamma \in \Lambda_n^{\#}/\Delta^{\#}\} \quad .$$

The goal of this section is to rederive the classical multiplication theory formulas including the addition formulas in terms of representation theory. As we shall show these formulas come about by decomposing various elements in $A_n[{}^0_0](\tau)$ in terms of this basis. The crucial fact is that we have by the proof of Theorem 9 an explicit procedure for writing elements of $A_n[{}^0_0](\tau)$ in terms of the given basis.

The case $n = 2$ is considered first and we rederive for this case the formulas of [6] on pages 97-103 and 230-243.

Let $\sigma = \begin{pmatrix}1 & 0\\ 0 & 2\end{pmatrix}$, $\Delta = \sigma^{-1}(\Gamma Z) = (1/2,0,0)\Gamma Z$ and $J = \sigma(H_1) = J(\Delta^{\#})$. Then

$$H_2 = J \oplus L(0,1/2,0)J \quad .$$

The basis for $A_2[{}^0_0](\tau)$ described by Theorem 7 is ψ_{20}, ψ_{21} given by

$$\psi_{20} = e^{2\pi i\cdot 2\cdot x}\, e^{\pi i\tau\cdot 2y^2}\, \theta[{}^0_0](2\zeta|2\tau)$$

$$\psi_{21} = L(0,1/2,0)\ \psi_{20} = e^{2\pi i.2.x}\, e^{\pi i\tau.2y^2}\, \theta[{}^1_0](2\zeta|2\tau)$$

where $\zeta = x + \tau y$.

For any function $F \in A_2[{}^0_0](\tau)$, there exist unique constants K_1, K_2 (depending on τ) such that $F = K_1\psi_{20} + K_2\psi_{21}$.

On the other hand using the proof of Theorem 9 we can also write

$$\text{(7)}\qquad \begin{aligned} K_1\psi_{20} &= p_J(F) = \tfrac{1}{2}(F + L(\tfrac{1}{2},0,0)\ F)\\ K_2\psi_{20} &= p_{J^\perp}(F) = p_J\ (L(0,\tfrac{1}{2},0)\ F) \end{aligned}$$

The following easily derived formula concerning the evaluation of theta functions at half periods will be useful in computing explicit formulas for K_1, K_2 below.

$$\theta\left[\begin{smallmatrix} m_1 \\ m_2 \end{smallmatrix}\right]\left(\zeta + \frac{u_2 + u_1\tau}{2} \,\middle|\, \tau\right) = e^{\pi i\left(-\frac{\tau u_1^2}{4} - \frac{u_1}{2}(m_2+u_2) - u_2\zeta\right)} \theta\left[\begin{smallmatrix} m_1+u_1 \\ m_2+u_2 \end{smallmatrix}\right](\zeta|\tau) \; . \tag{8}$$

<u>Theorem IV.11</u>: Let $G = \psi\left[\begin{smallmatrix} m_1 \\ m_2 \end{smallmatrix}\right](x,y,z|\tau)\,\psi\left[\begin{smallmatrix} -m_1 \\ -m_2 \end{smallmatrix}\right](x,y,z|\tau)$. Then $G \in A_2\left[\begin{smallmatrix} 0 \\ 0 \end{smallmatrix}\right](\tau)$.

Write $G = K_1\left[\begin{smallmatrix} m_2 \\ m_2 \end{smallmatrix}\right]\psi_{20} + K_2\left[\begin{smallmatrix} m_1 \\ m_2 \end{smallmatrix}\right]\psi_{21}$. Then

$$K_1\left[\begin{smallmatrix} m_1 \\ m_2 \end{smallmatrix}\right] = \tfrac{1}{2}\,\frac{\theta\left[\begin{smallmatrix} m_1 \\ m_2 \end{smallmatrix}\right](\zeta|\tau)\theta\left[\begin{smallmatrix} -m_1 \\ -m_2 \end{smallmatrix}\right](\zeta|\tau) + \theta\left[\begin{smallmatrix} m_1 \\ m_2+1 \end{smallmatrix}\right](\zeta|\tau)\theta\left[\begin{smallmatrix} -m_1 \\ -m_2+1 \end{smallmatrix}\right](\zeta|\tau)}{\theta\left[\begin{smallmatrix} 0 \\ 0 \end{smallmatrix}\right](2\zeta|2\tau)} \tag{9.a}$$

$$K_2\left[\begin{smallmatrix} m_1 \\ m_2 \end{smallmatrix}\right] = \tfrac{1}{2}\,\frac{\theta\left[\begin{smallmatrix} m_1+1 \\ m_2 \end{smallmatrix}\right](\zeta|\tau)\theta\left[\begin{smallmatrix} -m_1+1 \\ -m_2 \end{smallmatrix}\right](\zeta|\tau) + \theta\left[\begin{smallmatrix} m_1+1 \\ m_2+1 \end{smallmatrix}\right](\zeta|\tau)\theta\left[\begin{smallmatrix} -m_1+1 \\ -m_2+1 \end{smallmatrix}\right](\zeta|\tau)}{\theta\left[\begin{smallmatrix} 0 \\ 0 \end{smallmatrix}\right](2\zeta|2\tau)} \tag{9.b}$$

<u>Proof</u>: Clearly from (3)

$$L(\tfrac{1}{2},0,0)G = e^{-2\pi i\frac{1}{2}y}\, e^{-2\pi i\frac{1}{2}\frac{m_1}{2}}\, \psi\left[\begin{smallmatrix} m_1 \\ m_2+1 \end{smallmatrix}\right](x,y,z|\tau)\cdot e^{-2\pi i\frac{1}{2}y}\, e^{-2\pi i\left(\frac{-m_1}{2}\right)}\, \psi\left[\begin{smallmatrix} -m_1 \\ -m_2+1 \end{smallmatrix}\right](x,y,z|\tau)$$

$$= e^{-2\pi i y}\, \psi\left[\begin{smallmatrix} m_1 \\ m_2+1 \end{smallmatrix}\right](x,y,z|\tau)\, \psi\left[\begin{smallmatrix} -m_1 \\ -m_2+1 \end{smallmatrix}\right](x,y,z|\tau) \; .$$

Hence

$$K_1\, e^{2\pi i\, 2x}\, e^{\pi i\tau\, 2y^2}\, \theta\left[\begin{smallmatrix} 0 \\ 0 \end{smallmatrix}\right](2\zeta|2\tau)$$

$$= \tfrac{1}{2}\left(\psi\left[\begin{smallmatrix} m_1 \\ m_2 \end{smallmatrix}\right](x,y,z|\tau)\psi\left[\begin{smallmatrix} -m_1 \\ -m_2 \end{smallmatrix}\right](x,y,z|\tau) + e^{-2\pi i y}\, \psi\left[\begin{smallmatrix} m_1 \\ m_2+1 \end{smallmatrix}\right](x,y,z|\tau)\psi\left[\begin{smallmatrix} -m_1 \\ -m_2+1 \end{smallmatrix}\right](x,y,z|\tau\right.$$

$$= \tfrac{1}{2}\, e^{2\pi i\, 2x}\, e^{\pi i\tau\, 2y^2}\left(\theta\left[\begin{smallmatrix} m_1 \\ m_2 \end{smallmatrix}\right](\zeta|\tau)\theta\left[\begin{smallmatrix} -m_1 \\ -m_2 \end{smallmatrix}\right](\zeta|\tau) + \theta\left[\begin{smallmatrix} m_1 \\ m_2+1 \end{smallmatrix}\right](\zeta|\tau)\theta\left[\begin{smallmatrix} -m_1 \\ -m_2+1 \end{smallmatrix}\right](\zeta|\tau)\right).$$

The first statement of (9) follows.

The second statement of (9) follows from (7) in the same way completing the proof of the theorem.

Replacing ζ by $\zeta + \tau/2$ in (9) and using (8) gives the following formula.

$$K_2\left[\begin{smallmatrix} m_1 \\ m_2 \end{smallmatrix}\right] = \frac{1}{2}\ \frac{\theta\left[\begin{smallmatrix} m_1 \\ m_2 \end{smallmatrix}\right](\zeta|\tau)\theta\left[\begin{smallmatrix} -m_1 \\ -m_2 \end{smallmatrix}\right](\zeta|\tau) - \theta\left[\begin{smallmatrix} m_1 \\ m_2+1 \end{smallmatrix}\right](\zeta|\tau)\theta\left[\begin{smallmatrix} -m_1 \\ -m_2+1 \end{smallmatrix}\right](\zeta|\tau)}{\theta\left[\begin{smallmatrix} 1 \\ 0 \end{smallmatrix}\right](2\zeta|2\tau)} .$$

Corollary:

(10.a) $$\theta\left[\begin{smallmatrix} 0 \\ 0 \end{smallmatrix}\right](2\zeta|2\tau) = \frac{\theta\left[\begin{smallmatrix} 1 \\ 0 \end{smallmatrix}\right](0|2\tau)}{\theta^2\left[\begin{smallmatrix} 1 \\ 0 \end{smallmatrix}\right]}\left(\theta^2\left[\begin{smallmatrix} 0 \\ 0 \end{smallmatrix}\right](\zeta|\tau) + \theta^2\left[\begin{smallmatrix} 0 \\ 1 \end{smallmatrix}\right](\zeta|\tau)\right)$$

(10.b) $$\theta\left[\begin{smallmatrix} 1 \\ 0 \end{smallmatrix}\right](2\zeta|\tau) = \frac{\theta\left[\begin{smallmatrix} 1 \\ 0 \end{smallmatrix}\right](0|2\tau)}{\theta^2\left[\begin{smallmatrix} 1 \\ 0 \end{smallmatrix}\right]}\left(\theta^2\left[\begin{smallmatrix} 1 \\ 0 \end{smallmatrix}\right](\zeta|\tau) - \theta^2\left[\begin{smallmatrix} 0 \\ 1 \end{smallmatrix}\right](\zeta|\tau)\right) .$$

Proof: Statement (10.a) follows from (9.b) with $m_1 = 1$, $m_2 = 0$ the constant $K_2\left[\begin{smallmatrix} 1 \\ 0 \end{smallmatrix}\right]$ being evaluated at $\tau/2$.

Statement (10.b) follows from (9.b) with $m_1 = 1$, $m_2 = 0$ the constant $K_2\left[\begin{smallmatrix} 1 \\ 0 \end{smallmatrix}\right]$ being evaluated at 0 noting $\theta\left[\begin{smallmatrix} 1 \\ 1 \end{smallmatrix}\right](0|\tau) \equiv 0$.

The reader may wish to compare with (3) and (4) of page 231 of [6].

We shall now derive general formulas for the coefficients of arbitrary products of two first order theta functions. To do so we first remark that in general a product $\psi\left[\begin{smallmatrix} m_1 \\ m_2 \end{smallmatrix}\right](x,y,z|\tau)\psi\left[\begin{smallmatrix} n_1 \\ n_2 \end{smallmatrix}\right](x,y,z|\tau)$ does not lie in $A_2\left[\begin{smallmatrix} 0 \\ 0 \end{smallmatrix}\right](\tau)$ and hence can no longer be written as a linear combination with constant coefficients in ψ_{20}, ψ_{21} . In fact this product lies in $A_2\left[\begin{smallmatrix} m_1+n_1 \\ m_2+n_2 \end{smallmatrix}\right](\tau)$ and we must begin by determining a basis of this space.

Set

$$\psi_{20}\left[\begin{smallmatrix} m_1 \\ m_2 \end{smallmatrix}\right] = R(m_2/4, m_1/4, m_1 m_2/16)\psi_{20}$$

$$\psi_{21}\left[\begin{smallmatrix} m_1 \\ m_2 \end{smallmatrix}\right] = R(m_2/4, m_1/4, m_1 m_2/16)\psi_{21} .$$

A simple computation shows that

$$\psi_{20}\begin{bmatrix} m_1 \\ m_2 \end{bmatrix} = M_2\begin{bmatrix} m_1 \\ m_2 \end{bmatrix}(\tau)\ \theta\begin{bmatrix} m_1/2 \\ m_2 \end{bmatrix}(2\zeta|2\tau), \quad \zeta = x + \tau y$$

(11)

$$\psi_{21}\begin{bmatrix} m_1 \\ m_2 \end{bmatrix} = M_2\begin{bmatrix} m_1 \\ m_2 \end{bmatrix}(\tau)\ \theta\begin{bmatrix} m_1/2 + 1 \\ m_2 \end{bmatrix}(2\zeta|2\tau), \quad \zeta = x + \tau y$$

where $\theta\begin{bmatrix} m_1/2 \\ m_2 \end{bmatrix}(2\zeta|2\tau)$ and $\theta\begin{bmatrix} m_1/2 + 1 \\ m_2 \end{bmatrix}(2\zeta|2\tau)$ are second order analytic theta functions of period τ, argument ζ and characteristic $\begin{bmatrix} m_1 \\ m_2 \end{bmatrix}$.

The remark that is crucial is that $\psi_{20}\begin{bmatrix} m_1 \\ m_2 \end{bmatrix}$, $\psi_{21}\begin{bmatrix} m_1 \\ m_2 \end{bmatrix}$ give a basis of $A_2\begin{bmatrix} m_1 \\ m_2 \end{bmatrix}(\tau)$ so that $\psi_{20}\begin{bmatrix} m_1 \\ m_2 \end{bmatrix}$ generates $J_A\begin{bmatrix} m_1 \\ m_2 \end{bmatrix}(\tau)$ and $\psi_{21}\begin{bmatrix} m_1 \\ m_2 \end{bmatrix}$ generates $J_A^{\perp}\begin{bmatrix} m_1 \\ m_2 \end{bmatrix}(\tau)$. We use $J^{\perp}$ to mean the complement of J in H_2.

<u>Theorem IV.12</u>: Let $G = \psi\begin{bmatrix} m_1 \\ m_2 \end{bmatrix}(x,y,z|\tau)\psi\begin{bmatrix} n_1 \\ n_2 \end{bmatrix}(x,y,z|\tau)$. Then

$$G = K_1\begin{bmatrix} \frac{m_1-n_1}{2} \\ \frac{m_2-n_2}{2} \end{bmatrix} \psi_{20}\begin{bmatrix} m_1+n_1 \\ m_2+n_2 \end{bmatrix} + K_2\begin{bmatrix} \frac{m_1-n_1}{2} \\ \frac{m_2-n_2}{2} \end{bmatrix} \psi_{21}\begin{bmatrix} m_1+n_1 \\ m_2+n_2 \end{bmatrix}$$

where $K_1\begin{bmatrix} r_1 \\ r_2 \end{bmatrix}$, $K_2\begin{bmatrix} r_1 \\ r_2 \end{bmatrix}$ are given as in the preceeding theorem. Furthermore,

$$K_1\begin{bmatrix} \frac{m_1-n_1}{2} \\ \frac{m_2-n_2}{2} \end{bmatrix} = \frac{1}{2}\,\frac{\theta\begin{bmatrix} m_1 \\ m_2 \end{bmatrix}(\zeta|\tau)\theta\begin{bmatrix} n_1 \\ n_2 \end{bmatrix}(\zeta|\tau) + e^{-\pi i\left(\frac{m_1+n_1}{2}\right)}\theta\begin{bmatrix} m_1 \\ m_2+1 \end{bmatrix}(\zeta|\tau)\theta\begin{bmatrix} n_1 \\ n_2+1 \end{bmatrix}(\zeta|\tau)}{\theta\begin{bmatrix} \frac{m_1+n_1}{2} \\ m_2+n_2 \end{bmatrix}(2\zeta|2\tau)}$$

$$K_2\begin{bmatrix} \frac{m_1-n_1}{2} \\ \frac{m_2-n_2}{2} \end{bmatrix} = \frac{1}{2}\,\frac{\theta\begin{bmatrix} m_1+1 \\ m_2 \end{bmatrix}(\zeta|\tau)\theta\begin{bmatrix} n_1+1 \\ n_2 \end{bmatrix}(\zeta|\tau) + e^{-\pi i\left(\frac{m_1+n_1}{2}\right)}\theta\begin{bmatrix} m_1+1 \\ m_2+1 \end{bmatrix}(\zeta|\tau)\theta\begin{bmatrix} n_1+1 \\ n_2+1 \end{bmatrix}(\zeta|\tau)}{\theta\begin{bmatrix} \frac{m_1+n_1}{2} \\ m_2+n_2 \end{bmatrix}(2\zeta|2\tau)}$$

$$R\left(\frac{-(m_2+n_2)}{4}, 0, \frac{-(m_1+n_1)}{4}\right)G = \psi\begin{bmatrix}\frac{m_1-n_1}{2}\\ \frac{m_2-n_2}{2}\end{bmatrix}(x,y,z|\tau)\ \psi\begin{bmatrix}\frac{n_1-m_1}{2}\\ \frac{n_2-m_2}{2}\end{bmatrix}(x,y,z|\tau) ,$$

and then operating by $\left(\frac{m_2+n_2}{4}, \frac{m_1+n_1}{4}, \frac{(m_1+n_1)(m_2+n_2)}{16}\right)$ gives the first formula.

Operating by $\frac{1}{2}(I+L(\frac{1}{2},0,0))$ where I is the identity operator gives

$$K_1\begin{bmatrix}\frac{m_1-n_1}{2}\\ \frac{m_2-n_2}{2}\end{bmatrix}\psi_{20}$$

as before. Hence operating by $\frac{1}{2}(I+L(\frac{1}{2},0,0))$ on G gives

$$K_1\begin{bmatrix}\frac{m_1-n_1}{2}\\ \frac{m_2-n_2}{2}\end{bmatrix}\ \psi_{20}\begin{bmatrix}m_1+n_1\\ m_2+n_2\end{bmatrix} .$$

The second statement follows after using (4) to express $\frac{1}{2}(I+L(\frac{1}{2},0,0))G$, and cancelling. The final statement follows as well.

Replacing ζ by $\zeta + \tau/2$ and using (8) gives

$$K_2\begin{bmatrix}\frac{m_1-n_1}{2}\\ \frac{m_2-n_2}{2}\end{bmatrix} = \frac{1}{2}\,\frac{\theta[{}^{m_1}_{m_2}](\zeta|\tau)\theta[{}^{n_1}_{n_2}](\zeta|\tau) - e^{-\pi i(\frac{m_1+n_1}{2})}\,\theta[{}^{m_1}_{m_2+1}](\zeta|\tau)\theta[{}^{n_1}_{n_2+1}](\zeta|\tau)}{\theta\begin{bmatrix}\frac{m_1+n_1}{2}+1\\ m_2+n_2\end{bmatrix}(2\zeta|2\tau)} .$$

<u>Corollary</u>:

$$\theta[{}^1_1](2\zeta|2\tau) = \frac{\theta[{}^0_1](0|2\tau)}{\theta[{}^0_1]\theta[{}^0_0]}\ \theta[{}^1_1](\zeta|\tau)\theta[{}^1_0](\zeta|\tau)$$

$$\theta[{}^0_1](2\zeta|2\tau) = \frac{\theta[{}^0_1](0|2\tau)}{\theta[{}^0_1]\theta[{}^0_0]} \quad \theta[{}^0_1](\zeta|\tau)\theta[{}^0_0](\zeta|\tau) \quad .$$

Note this corollary is also found on page 231 of [6].

We come now to the derivation of the theta identities and addition formulas (see pages 94-103 of [6]) in terms of our formulation.

For $m = (m_1,m_2,m_3,m_4) \in \mathbb{R}^4$ set

$$\psi[{}^{m_1}_{m_2}] \cdot \psi[{}^{m_3}_{m_4}] = K_0(m)\psi_{20}(m) + K_1(m)\psi_{21}(m) \ .$$

$$\psi_{2i}(m) = \psi_{2i}[{}^{m_1+m_2}_{m_2+m_4}] \quad , \quad i = 0,1.$$

$$K_i(m) = K_i \begin{bmatrix} \frac{m_1-m_2}{2} \\ \frac{m_2-m_4}{2} \end{bmatrix} \quad , \quad i = 0,1 \quad .$$

For $m, \ell, r \in \mathbb{R}^4$ set

$$C(m;\ell,r) = \frac{K_1(m)K_2(r) - K_1(r)K_2(m)}{K_1(\ell)K_2(r) - K_1(r)K_2(\ell)} \quad .$$

A direct computation using Theorem 12 shows that we can write $C(m;n,r)$ as follows.

$$C(m;\ell,r) = \frac{C(m;r) - C(r;m)}{C(\ell;r) - C(r;\ell)}$$

where

$$C(m;r) = \theta[{}^{m_1}_{m_2+1}](\zeta|\tau)\theta[{}^{m_3}_{m_4+1}](\zeta|\tau)\theta[{}^{r_1}_{r_2}](\zeta|\tau)\theta[{}^{r_3}_{r_4}](\zeta|\tau) \ .$$

Note that $C(m;r)$ is a function of ζ and τ and $C(m;\ell,r)$ is a constant (depending solely on τ) where the order of ℓ and r must be taken into consideration.

Theorem IV.13: For $m, \ell, r \in \mathbb{R}^4$ satisfying $m_1 + m_3 = \ell_1 + \ell_3 = r_1 + r_3, m_2 + m_4 = \ell_2 + \ell_4 = r_2 + r_4$,

$$G_m = C(m;\ell,r)G_\ell + C(m;r,\ell)G_r .$$

Proof: By Theorem 12

$$G_m = K_1(m)\,\psi_{20}(m) + K_2(m)\,\psi_{21}(m)$$

$$G_r = K_1(\ell)\,\psi_{20}(\ell) + K_2(\ell)\,\psi_{21}(\ell)$$

$$G_r = K_1(r)\,\psi_{20}(r) + K_2(r)\,\psi_{21}(r) .$$

Hence using our assumption $\psi_{2i}(m) = \psi_{2i}(r)$ for $i = 0,1$, we can write $\psi^*_{2i} = \psi_{2i}(m)$, and then

$$\begin{pmatrix} \psi^*_{20} \\ \psi^*_{21} \end{pmatrix} = \frac{1}{K_1(\ell)K_2(r) - K_2(\ell)K_1(r)} \begin{pmatrix} K_2(r) & -K_2(\ell) \\ -K_1(r) & K_1(\ell) \end{pmatrix} \begin{pmatrix} G_\ell \\ G_r \end{pmatrix}$$

It follows that G_m is given as stated.

Corollary:

(1) $$-\theta^2\left[\begin{smallmatrix}0\\0\end{smallmatrix}\right]\theta^2\left[\begin{smallmatrix}0\\0\end{smallmatrix}\right](\zeta|\tau) + \theta^2\left[\begin{smallmatrix}1\\0\end{smallmatrix}\right]\theta^2\left[\begin{smallmatrix}1\\0\end{smallmatrix}\right](\zeta|\tau) + \theta^2\left[\begin{smallmatrix}0\\1\end{smallmatrix}\right]\theta^2\left[\begin{smallmatrix}0\\1\end{smallmatrix}\right](\zeta|\tau) \equiv 0 .$$

(2) $$\theta^2\left[\begin{smallmatrix}0\\0\end{smallmatrix}\right]\theta^2\left[\begin{smallmatrix}1\\1\end{smallmatrix}\right](\zeta|\tau) - \theta^2\left[\begin{smallmatrix}1\\0\end{smallmatrix}\right]\theta^2\left[\begin{smallmatrix}0\\1\end{smallmatrix}\right](\zeta|\tau) + \theta^2\left[\begin{smallmatrix}0\\1\end{smallmatrix}\right]\theta^2\left[\begin{smallmatrix}1\\0\end{smallmatrix}\right](\zeta|\tau) \equiv 0 .$$

(3) $$-\theta^2\left[\begin{smallmatrix}0\\0\end{smallmatrix}\right]\theta^2\left[\begin{smallmatrix}0\\1\end{smallmatrix}\right](\zeta|\tau) + \theta^2\left[\begin{smallmatrix}1\\0\end{smallmatrix}\right]\theta^2\left[\begin{smallmatrix}1\\1\end{smallmatrix}\right](\zeta|\tau) + \theta^2\left[\begin{smallmatrix}0\\1\end{smallmatrix}\right]\theta^2\left[\begin{smallmatrix}0\\0\end{smallmatrix}\right](\zeta|\tau) \equiv 0 .$$

(4) $$\theta^2\left[\begin{smallmatrix}0\\0\end{smallmatrix}\right]\theta^2\left[\begin{smallmatrix}1\\0\end{smallmatrix}\right](\zeta|\tau) - \theta^2\left[\begin{smallmatrix}1\\0\end{smallmatrix}\right]\theta^2\left[\begin{smallmatrix}0\\0\end{smallmatrix}\right](\zeta|\tau) + \theta^2\left[\begin{smallmatrix}0\\1\end{smallmatrix}\right]\theta^2\left[\begin{smallmatrix}1\\1\end{smallmatrix}\right](\zeta|\tau) \equiv 0 .$$

Take $m, \ell, r \in \mathbb{R}^4$ with $m_1 + m_3 = \ell_1 + \ell_3 = r_1 + r_3$ and $m_2 + m_4 = \ell_2 + \ell_4 = r_2 + r_4$.

For $\gamma = (x,y,z) \in N$ define the function $G_{m,\eta}$ on N by

$$G_{m,\eta}(\delta|\tau) = \psi\left[\begin{smallmatrix}m_1\\m_2\end{smallmatrix}\right](\delta\gamma|\tau)\;\psi\left[\begin{smallmatrix}m_3\\m_4\end{smallmatrix}\right](\delta\gamma^{-1}|\tau) , \quad \delta \in N$$

where $\eta = x + \tau y$.

It is easy to compute that this definition is well defined and arguing as in Theorem 12 we see that

$$G_{m,\eta}(\delta|\tau) = K_1(m,\eta)\ \psi_{20}(m)(\delta|\tau) + K_2(m,\eta)\ \psi_{21}(m)(\delta|\tau)$$

where the $K_i(m,\eta)$ are uniquely determined constants depending only on m,η and τ.

As before

$$G_\ell = K_1(\ell)\ \psi_{20}^* + K_2(\ell)\ \psi_{21}^*$$

$$G_r = K_1(r)\ \psi_{20}^* + K_2(r)\ \psi_{21}^*$$

with $\psi_{2i}^* = \psi_{2i}(m)$.

Clearly

$$g_{m,\eta}(\delta|\tau) = \frac{K_1(m,\eta)K_2(r) - K_2(m,\eta)K_1(r)}{K_1(\ell)K_2(r) - K_2(\ell)K_1(r)}\ G_\ell(\delta|\tau)$$

$$+ \frac{K_2(m,\eta)K_1(\ell) - K_1(m,\eta)K_2(n)}{K_1(\ell)K_2(r) - K_2(\ell)K_1(r)}\ G_r(\delta|\tau) .$$

Hence setting

$$C(m,\eta;\ell,r) = \frac{C(m,\eta;r) - C(r,\eta;m)}{C(\ell,r) - C(r,\ell)}$$

where

$$C(m,\eta;r) = \theta[{}^{m_1}_{m_2+1}](\zeta+\eta|\tau)\theta[{}^{m_3}_{m_4+1}](\zeta-\eta|\tau)\theta[{}^{r_1}_{r_2}](\zeta|\tau)\theta[{}^{r_3}_{r_4}](\zeta|\tau)$$

it follows as before that

$$G_{m,\eta}(\delta|\tau) = C(m,\eta;\ell,r)G_\ell + C(m,\eta;r,\ell)G_r .$$

This last formula includes the addition theorems of [6], i.e., (19), (20),

For example we prove the following.

<u>Corollary</u>:

$$\theta^2[{}^0_1]\theta[{}^0_1](\zeta+\eta|\tau)\theta[{}^0_1](\zeta-\eta|\tau) = \theta^2[{}^0_1](\zeta|\tau)\theta^2[{}^0_1](\eta|\tau) - \theta^2[{}^1_1](\zeta|\tau)\theta^2[{}^1_1](\eta|\tau).$$

<u>Proof</u>: Set $m_1 = m_3 = 0$, $m_2 = m_4 = 1$, $\ell_1 = \ell_3 = 0$, $\ell_2 = \ell_4 = 1$, $r_1 = 1$, $r_3 = -1$, $r_2 = r_4 = 1$. The statement of the corollary immediately follows.

We close by investigating some of the ideas above for the case n an odd but otherwise arbitrary integer. In particular we rederive in our language the formulas of page 239-241 of [6].

Let n be an odd integer. Set $J = J(\Delta) = \sigma = \begin{pmatrix} 1 & 0 \\ 0 & n \end{pmatrix}$, $\Delta = \sigma^{-1}(\Gamma Z) = (1/n,0,0)\Gamma Z$. Then specializing from our previous results we have

$$H_n = \sum_{j=0}^{n-1} \oplus\, L(0,\tfrac{-j}{n},0)J$$

$$A_n[{}^0_0](\tau) = \sum_{j=0}^{n-1} \oplus\, L(0,-j/n,0)\; J_A[{}^0_0](\tau)$$

and the elements

$$\psi_{n0} = e^{2\pi inx}\, e^{\pi i\tau ny^2}\, \theta[{}^0_0](n\zeta|n\tau)$$

$$\psi_{nj} = L(0,-j/n,0)\; \psi_{n0} = e^{2\pi inx}\, e^{\pi i\tau ny^2}\, \theta[{}^{2j/n}_{0}](n\zeta|n\tau)$$

define a basis of $A_n[{}^0_0](\tau)$ with ψ_{nj} generating the one dimensional complex vector space $L(0,-j/n,0)\; J_A[{}^0_0](\tau)$, $j = 0,\ldots,n-1$.

For any $F \in A_n[{}^0_0](\tau)$ we can write

$$F = K_0\, \psi_{n0} +,\ldots,+ K_{n-1}\, \psi_{n,n-1}$$

where the constants K_j are given by the rule (different K's from above)

$$K_j = \frac{1+L(1/n,0,0) +,\ldots,+ L(n-1/n,0,0)\; L(0,j/n,0)F}{\psi_{n0}} .$$

Set

$$\psi^{\#}_{n0} = \psi_{n0}$$

$$\psi^{\#}_{nj} = \psi_{nj} + \psi_{n,n-j}\,, \quad j =,\ldots,m = \frac{n-1}{2} .$$

The elements $\psi^{\#}_{nj}$, $j=0,\ldots,m$ clearly define a basis for the even functions in $A_n[{}^0_0](\tau)$, i.e., the image of the even functions of $\Theta_n[{}^0_0](\tau,A)$ in $A_n[{}^0_0](\tau)$. For any even function $F \in A_n[{}^0_0](\tau)$ the formula for K_j immediately implies

$$K_j = K_{n-j}\,, \quad j = 1,\ldots,m .$$

Hence

$$F = K_0 \psi^{\#}_{n0} + K_1 \psi^{\#}_{n1} + \dots + K_m \psi^{\#}_{nm}$$

Set

$$\Psi_0 = \psi^n[{}^0_0](x,t,z|\tau) \ , \ \Psi_K = \psi^{n-2k}[{}^0_0](x,y,z|\tau)\ \psi^{2k}[{}^1_1](x,y,z|\tau) \ .$$

A simple computation shows that

$$\Psi_k = \sum_{j=0}^{m} K_{jk}\, \psi^{\#}_{nj}$$

where

$$\theta[{}^0_0](0|n\tau)\cdot K_{jk} = \sum_{\ell=0}^{m} \theta^{n-2k}[{}^{-2j/n}_{0}](\tfrac{\ell}{n}|\tau)\theta^{2k}[{}^{1-2j/n}_{1}](\tfrac{\ell}{n}|\tau) \ .$$

Finally we must show that the matrix $K = (K_{jk})$ is invertible and compute $K^*_{01},\dots,K^*_{0m}$ the first row of its inverse. A direct computation which we leave to the reader shows that K is invertible and

$$\theta^n[{}^0_0]^{-1}\ \theta[{}^0_0](0|2\tau)\cdot K^*_{00} = 1$$

$$\theta^n[{}^0_0]\theta^{-1}[{}^0_0](0|2\tau)\ K^*_{10} = -\sum_{j=1}^{m} \frac{\theta^2[{}^0_0](\frac{j}{n}|\tau)}{\theta^2[{}^1_1](\frac{j}{n}|\tau)}$$

$$\vdots$$

$$\theta^n[{}^0_0]\theta^{-1}[{}^0_0](0|2\tau)\ K^*_{m0} = \prod_{j=1}^{m} \frac{\theta^2[{}^0_0](\frac{j}{n}|\tau)}{\theta^2[{}^1_1](\frac{j}{n}|\tau)} \ .$$

We state our final result as a theorem.

Theorem IV.14: For odd integer $n > 0$

$$\theta[{}^0_0](n\zeta|n\tau) = \frac{\theta[{}^0_0](0|n\tau)}{\theta^n[{}^0_0]}\ \theta[{}^0_0](\zeta|\tau)\cdot \prod_{\ell=1}$$

$$\times \frac{(\theta^2[{}^0_0](\zeta|\tau)\ \theta^2[{}^1_1](\frac{\ell}{n}|\tau) + (-1)\theta^2[{}^1_1](\zeta|\tau)\ \theta^2[{}^0_0](\frac{\ell}{n}|\tau))}{\theta^2[{}^1_1](\frac{\ell}{n}|\tau)} \ .$$

The other formulas of [6] on the pages 239-241 result from half period substitutions.

APPENDIX B

THE ARITHMETIC OF DISTINGUISHED SUBSPACES

We recall the following notation. For $n \geq 1$, $\mathcal{D}_n$ equals the collection of all distinguished spaces on H_n. Assume for simplicity that $n \equiv 0 \bmod 2$ in the remainder of this appendix. In this case recall that $SL_2(Z)$ acts as a group of permutations of $\mathcal{D}_n$. The first object of this appendix is to describe and count the collection $\mathcal{D}_n$ and the space of orbits $\mathcal{D}_n/SL_2(Z)$. The techniques are completely algebraic and generally speaking the numbers we come up with will be arithmetically related to n

For $J \in \mathcal{D}_n$ the isotopy subgroup $I(J)$ of J is defined by

$$I(J) = \{\sigma \in SL_2(Z)\colon \sigma(J) = J\} .$$

The second object of this appendix is to describe the group $I(J)$.

We begin this study by recalling certain facts from Chapter IV. As we shall see, these problems, although stated as problems involving the Heisenberg group, are equivalent to problems involving the groups $S_n = Z/nZ \oplus Z/nZ$ previously considered in Chapter IV.

Setting, as before, $\mathcal{E}_n$ equal to the collection of all subgroups Δ of S_n of order n and $ch(\Delta)$ equal to the group of characters of Δ, the results of Chapter IV establish a bijection between the collection $\mathcal{D}_n$ and the collection of all pairs (Δ,λ), where $\Delta \in \mathcal{E}_n$ and $\lambda \in ch(\Delta)$. We abbreviate this bijection by writing $J = (\Delta,\lambda)$.

The action of $SL_2(Z)$ on $\mathcal{D}_n$ corresponds under the above bijection to the action on the pairs (Δ,λ) induced by the action of $SL_2(Z)$ on S_n given by $\sigma(u,v) = (ay+bv, cu+dv)$, where $\sigma = \begin{pmatrix} a & b \\ c & d \end{pmatrix} \in SL_2(Z)$ and $(u,v) \in S_n$. Hence if $J = (\Delta,\lambda)$ and $\sigma \in SL_2(Z)$ then $\sigma(J) = (\sigma^{-1}(\Delta), \lambda \circ \sigma)$.

This last formula implies that for any $J = (\Delta,\lambda) \in \mathcal{D}_n$ the isotopy subgroup $I(J)$ is given by

$$I(J) = \{\sigma \in SL_2(Z)\colon \sigma^{-1}(\Delta) = \Delta,\ \lambda \circ \sigma = \lambda\} .$$

Also for $\sigma \in SL_2(Z)$ we have $\sigma^{-1}I(J)\sigma = I(\sigma(J))$. Let $\mathcal{J}_n$ denote the collection of all isotopy subgroups corresponding to the elements of $\mathcal{D}_n$. We see by the above remarks that $\mathcal{J}_n$ is closed under conjugation by the elements in $SL_2(Z)$. The corresponding orbit space we denote by $\mathcal{J}_n/SL_2(Z)$.

We are now ready to begin counting. Assume first that n is the <u>product of distinct primes</u>. In this case the answer we get will be explicit in contrast to the general case. Later we shall see what can be said about arbitrary n .

As proved in Chapter IV, the groups

$$\Lambda(n,m) = (m,0)^{\wedge} \oplus (0,\tfrac{n}{m})^{\wedge}, \quad m^2/n$$

determine a complete system of representatives for $\mathcal{E}_n/SL_2(Z)$. Since 1 is the only integer whose square divides n, $\mathcal{E}_n/SL_2(Z)$ consists of a single element which we take to be $\Lambda(n,1) = (1,0)$.

Let $\pi(n) = o(\mathcal{D}_n)$ and $\delta(n) = o(\mathcal{D}_n/SL_2(Z))$,where given a finite set X, $o(X)$ = the number of elements in X .

<u>Theorem B.1</u>: For $n \equiv 0 \bmod 2$ a product of distinct primes we have

1) $\pi(n) = n^2 \, \pi_{p/n}(1 + p^{-1})$.

2) $\delta(n)$ = the number of divisors of n including 1 and n itself.

<u>Proof</u>: Since $o(\mathcal{E}_n) = o(SL_2(Z)/G)$, where G is the subgroup of $SL_2(Z)$ defined by $G = \{\begin{pmatrix} a & b \\ c & d \end{pmatrix} \in SL_2(Z)\colon c \equiv 0 \bmod n\}$, we can use a result of [5] page 24 to get that $o(\mathcal{E}_n) = n\pi_{p/n}(1+p^{-1})$. Statement 1) follows since $o(ch(\Lambda(n,1)) = n$.

To compute $\delta(n)$ we first remark that $ch(\Lambda(n,1))$ consists of the elements λ_j, $0 \le j < n$ where λ_j is determined by the formula $\lambda_j(1,0) = e^{2\pi i \frac{j}{n}}$. Then $(\Lambda(n,1),\lambda_j)$ and $(\Lambda(n,1),\lambda_k)$ are in the same $SL_2(Z)$ orbit if and only if there exists a $\sigma \in G$ such that $\lambda_j \circ \sigma = \lambda_k$. This last condition is equivalent to the statement that $aj \equiv k \bmod n$ where $\sigma = \begin{pmatrix} a & b \\ c & d \end{pmatrix}$.

Take $0 \le j < n$. Set $\ell = G\,C\,D\,(j,n)$ if $j \ne 0$ and $\ell = n$ otherwise. Then since $G\,C\,D(n,\frac{j}{n} + x\,\frac{n}{\ell}) = 1$ for some integer x there exists a matrix

$$\sigma = \begin{pmatrix} r & s \\ n & \frac{j}{\ell} + x\frac{n}{\ell} \end{pmatrix} \in G$$

for some integers r,s. Using this matrix it is trivial to verify that $(\Lambda(n,1),\lambda_j)$ and $(\Lambda(n,1),\lambda_\ell)$ are in the same $SL_2(Z)$ orbit. Since ℓ/n every $J \in \mathcal{D}_n$ is $SL_2(Z)$ equivalent; ie., lies on the same $SL_2(Z)$-orbit as $(\Lambda(n,1),\lambda_\ell)$, where ℓ/n. It is easy to see that no two such $\mathcal{D}_n$'s are $SL_2(Z)$ equivalent and hence the collection $(\Lambda(n,1),\lambda_\ell),\ell/n$ determines a complete system of representatives of $\mathcal{D}_n/SL_2(Z)$. Statement 2) follows.

Before we begin the problem of describing the collection $\mathcal{J}_n$ we shall introduce some further definitions. Set $I = SL_2(Z)$. We shall first define some important subgroups of I. For $n > 1$ define

$$I_n = \{\begin{pmatrix} a & b \\ c & d \end{pmatrix} \in I\colon a \equiv d \equiv 1 \bmod n, c \equiv b \equiv 0 \bmod n\}.$$

I_n is called the principle congruence subgroup of level n and any subgroup $G \subseteq I$ containing some I_n is called a congruence subgroup. For our purposes it is necessary to define at this time one further kind of subgroup. For n,m both > 1 define

$$I(n,m) = \{\begin{pmatrix} a & b \\ c & d \end{pmatrix} \in I\colon a \equiv d \equiv 1 \bmod m,\ c \equiv 0 \bmod n\}.$$

The group G in the preceeding theorem is easily seen to be $I(n,1)$.

<u>Theorem B.2</u>: Let $n > 1$ be a product of distinct primes including 2. For $m > 1, m/\frac{n}{2}$ we have

$$o(I/I(n,m)) = n \cdot \varphi(m) \prod_{p/n} (1+p^{-1}),$$

where φ is the Euler φ-function.

<u>Proof</u>: This result must be classical but lacking a reference we indicate its proof. For $m = 1$ the result is found in [5]. Assume $m > 1$ then $O(I/I_m) = m^3 \prod_{p/m}(1-p^{-2})$ can be found in [5] page 24. It is easy to see that

$I_m \subset I(n,m)$ is a subgroup of order m with coset representatives given by $\begin{pmatrix} 1 & 0 \\ j & 1 \end{pmatrix}$, $j = 0,\ldots,m-1$. It remains to show $o(I(n,m)/I(n,m)) = \Pi_{j=1}^{r}(1+p_j)$ where $n = p_1 \ldots p_r\, m$ and the p_j's are prime. This fact can be verified one prime at a time; i.e., induction on r. The details are left to the reader.

<u>Theorem B.3</u>: The groups $I(n,m), m/\frac{n}{2}$ determine a complete system of representatives for $\mathcal{J}_n/I$. In particular

$$o(\mathcal{J}_n/I) = \frac{\delta(n)}{2} .$$

<u>Proof</u>: The group $I(n,\frac{n}{G\ C\,D(n,m)})$ is the isotopy subgroup of $J = (\Lambda(n,1),\lambda_m)$. Hence every $G \in \mathcal{J}_n$ is $SL_2(Z)$ equivalent to some $I(n,m)$ with m/n. Since $I(n,m) = I(n,2m)$ whenever $m/\frac{n}{2}$ it follows that we may restrict our attention to $I(n,m)$ with $m/\frac{n}{2}$.

Suppose $I(n,m)$ and $I(n,m')$, $m/\frac{n}{2}$, $m'/\frac{n}{2}$ are $SL_2(Z)$ equivalent. From the previous theorem a simple argument shows $m = m'$. The last statement of the theorem is now trivial and the theorem is proved.

Finally we comment that each $G \in \mathcal{J}_n$ is a congruent subgroup containing I_n and

$$I_n = \cap\, \mathcal{J}_n .$$

In the case of arbitrary $n \equiv 0 \bmod 2$ we do not get such an explicit form when we try to count. We shall briefly indicate the results with some indication at times of proof.

The groups $\Lambda(n,\ell)$, ℓ^2/n determine a complete system of representatives of $\mathcal{E}_n/I$.

Fix ℓ^2/n. Then the elements $\lambda^{(\ell)}_{j,k}$, $0 \le j < \frac{n}{\ell}$, $0 \le k < \ell$ defined by the formulas

$$\lambda^{(\ell)}_{j,k}(\ell,0) = e^{2\pi i \frac{j\ell}{n}}$$

$$\lambda^{(\ell)}_{j,k}(0,\tfrac{n}{\ell}) = e^{2\pi i \frac{k}{\ell}}$$

give the group $ch(\Lambda(n,\ell))$.

Although the pairs $(\Lambda(n,\ell),\lambda^{(\ell)}_{j,k}),\ell^2/n$ determine up to I equivalent $\mathcal{D}_n$, they do not lie on distinct orbits. In general the exact situation seems to be a difficult problem. However we do have the following.

Theorem B.4: $\pi(n) = n^2 \sum_{\ell^2/n} \frac{1}{\ell^2} \prod_{p/(n/\ell^2)} (1+p^{-1})$.

Proof: Follow directly from the classical result

$$o(I/I(\tfrac{n}{\ell^2},1)) = n/\ell^2 \cdot \prod_{p/\frac{n}{\ell^2}} (1+p^{-1}) .$$

For ℓ^2/n define

$w(n,\ell)$ = the number of I-orbits in $\mathcal{D}_n$ having $\Lambda(n,\ell)$ as stability group up to I equivalence.

This definition makes sense since any two I equivalent elements in $\mathcal{D}_n$ have I equivalent stability groups. Then

$$\delta(n) = \sum_{\ell^2/n} w(n,\ell) .$$

Now $S_{n,\ell} = Z/\frac{n}{\ell}Z \oplus Z/\ell Z$. We begin by defining the following equivalence relation: Two elements (j,k) and (j',k') are said to be $I(\frac{n}{\ell^2},1)$ equivalent if there exists a matrix

$$\sigma = \begin{pmatrix} a & b \\ c & d \end{pmatrix} \in I(\tfrac{n}{\ell^2},1)$$

such that

$$\text{(p)} \qquad (j,k) \equiv (j',k')\begin{pmatrix} a & b \\ c & d \end{pmatrix} \bmod \tfrac{n}{\ell} \bmod \ell .$$

It is easy to verify that two elements

$$(\Lambda(n,\ell),\lambda^{(\ell)}_{j,k}) \text{ and } (\Lambda(n,\ell),\lambda^{(\ell)}_{j',k'})$$

are I equivalent on $\mathcal{D}_n$ if and only if the pairs (j,k) and (j',k') are

$I(\frac{n}{\ell^2}, 1)$ equivalent. As a special case one can prove that if $GCD(\ell, \frac{n}{\ell^2}) = 1$ then $w(n,\ell) = \delta(\frac{n}{\ell})$.

Finally, we have some remarks about $\mathcal{J}_n$. For $n = \ell^2 u$ and integers $0 \le j < \ell u$, $0 \le k < \ell$ set

$$I(n,\ell_{jj},k) = \{ \begin{pmatrix} a & b \\ c & d \end{pmatrix} \in I(u,1) : (j,k)\begin{pmatrix} a & b \\ c & d \end{pmatrix} \equiv (j,k) \bmod \ell u \bmod \ell \} .$$

Then $I(n,\ell_{jj},k)$ is the isotopy subgroup of $(\Lambda(n,\ell), \lambda^{(\ell)}_{j,k})$. We state without proof our final result.

<u>Theorem B.5</u>: The groups $I(n,\ell_{jj},k)$, ℓ^2/n and $0 \le j < \frac{n}{\ell}$, $0 \le k < \ell$ chosen from a complete system of representatives relative to $I(\frac{n}{\ell^2}, 1)$ equivalence determines a complete system of representatives for $\mathcal{J}_n/I$.

As a final remark note that Theorem B.5 gives as a consequence that each $G \in \mathcal{J}_n$ is a congruence subgroup containing I_n and $I_n = \cap\, \mathcal{J}_n$.

REFERENCES

[1] L. Auslander and J. Brezin. Translation invariant subspaces in L^2 of a compact nilmanifold I, Inventiones Math. 20(1973), p. 1-14.

[2] J. Brezin. Function theory on metabelian solvmanifolds, J. of Func. Anal. 10(1972), p. 33-35.

[3] P. Cartier. Quantum mechanical commutative relations and theta functions, Proc. Symposia in Pure Math. IX, Amer. Math. Soc.(1965), p. 361-387.

[4] J. Igusa. Notes on theta functions, Institute for Advanced Study.

[5] S. Lang. Algebraic Number Theory, Addison-Wesley, (1970).

[6] H.E. Rauch and A. Lebowitz. Elliptic Functions, Theta Functions, and Riemann Surfaces, Williams and Wilkins, (1973).

[7] A. Weil. Sur certaines groupes d'operateurs unitaires, Acta. Math. 111(1964), p. 143-211.

[8] A. Weil. L'Integration dans les Groupes et ses applications, Topologique, Hermann, (1953).

APPENDIX C

FOURIER ANALYSIS ON THE HEISENBERG MANIFOLD

Until now in these notes we have stressed the interplay between the group theory and the representation theory associated with the nilmanifold $\Gamma\backslash N$ and the theory of theta functions. In essence, we have been justifying the contention that continuous theta functions can be viewed as the spherical harmonics of the nilmanifold $\Gamma\backslash N$. However, in Chapter II we gave strong indications that the theta functions might be used to study the spaces of smooth functions on $\Gamma\backslash N$, or what we will call the Fourier analysis on $\Gamma\backslash N$. It is to this topic that we will now turn our attention. Since our results in this direction are still tentative, in that we do not have a complete picture to present, but just some semi-isolated results, we have, following the convention of these notes, labelled this material an appendix.

Our basic point of view in this appendix is that $\Gamma\backslash N$ is a principal circle bundle over a torus T^2 and that the theta functions are the analytic expression of the cocycle defining this bundle. Thus we feel that the Fourier analysis of $\Gamma\backslash N$ should be built from the Fourier analysis of the torus T^2 and the fiber T^1 and the theta functions. The essential difficulty with this approach to the study of smooth functions on $\Gamma\backslash N$ is the fact that theta functions vanish on $\Gamma\backslash N$ as was discussed in Chapter II.

Because of the tentative nature of our results we have limited our study to smooth functions in H_1 and not even stated, using the techniques already presented in these notes, how these results may be extended to smooth functions in H_n. Similarly, we have limited our statements about smooth functions in H_1 to a few examples that illustrate how the Fourier analysis of the torus may be carried over to Fourier analysis in H_1. It should already have been clear that the decomposition

$$L^2(\Gamma\backslash N) = \oplus \sum_{n\in\mathbb{Z}} H_n$$

represents the role of the Fourier analysis of the fiber, or the circle, in our problem.

We have adopted two lines of approach to this problem, each stemming from a result, Theorems II.4 and II.5, in Chapter II. The first approach centers on the implications of the fact that $\psi_{10}e^{2\pi i(ax+by)}$, $a,b \in \mathbb{Z}$ forms a Hilbert space basis of H_1. The second centers about the fact that for $F \in C_1^r$, we can write,

$$F = p\psi_{10} + q\psi_{10}^* \quad , \quad p,q \in C^r(T^2)$$

in many ways.

<u>First Approach</u>: Let us now adopt the simpler notation $\psi = \psi_{10}$; $\psi^* = \psi_{10}^*$ and $\psi\chi(a,b) = \psi \exp 2\pi i(ax+by)$, $a,b \in \mathbb{Z}$. Recall then that $\psi\chi(a,b)$, $a,b \in \mathbb{Z}$, is dense in H_1 and no proper subset has this property. Unfortunately, the functions $\psi\chi(a,b)$ are not orthonormal in H^1 as shown by the following result.

<u>Lemma C.1</u>: Let (u_1,u_2), $(v_1,v_2) \in \mathbb{Z}^2$ and let $(\ ,\ )$ denote inner product in H_1. Then

$$(\psi\chi(u_1,u_2), \psi\chi(v_1,v_2)) = \frac{\sqrt{2}}{2} e^{-\frac{\pi}{2}((u_1-v_1)^2+(u_2-v_2)^2)} e^{\pi i(u_1-v_1)(u_2-v_2)} .$$

<u>Proof</u>: The function $\psi\chi(u_1,u_2)$ corresponds under the inverse of the Weil mapping to the function

$$e^{-\pi(t-u_2)^2} e^{2\pi i u_1 t} \in L^2(\mathbb{R})$$

where the Weil mapping is the intertwining operator $\mathcal{J}\Theta$ of Chapter I. Since $\mathcal{J}\Theta$ is a unitary operator between $L^2(\mathbb{R})$ and H_1 we have

$$(\psi\chi(u_1,u_2)\psi\chi(v_1,v_2)) = \int_{-\infty}^{\infty} e^{-\pi(t-u_2)^2} e^{-\pi(t-v_2)^2} e^{2\pi i(u_1-v_1)t} dt .$$

Setting $s = t-u_2$, we have

$$= \int_{-\infty}^{\infty} e^{-\pi s^2} e^{-\pi(s+u_2-v_2)^2} e^{2\pi i(u_1-v_1)s} ds$$

$$= e^{-\pi(u_2-v_2)^2} \int_{-\infty}^{\infty} e^{-2\pi s^2} e^{2s(-\pi(u_2-v_2)+\pi i(u_1-v_1))} ds .$$

Now by a classical analytic result, see Bellman [B] page 10, we may evaluate the above integral and obtain

$$(\psi\chi(u_1,u_2),\psi\chi(v_1,v_2) = \frac{\sqrt{2}}{2} e^{-\pi/2[(u_2-v_2)^2+(u_1-v_1)^2]} e^{\pi i[(u_1-v_1)+(u_2-v_2)]}.$$

Before examining further the implications of Lemma C.1, let us wee how we can use the function's $\psi\chi(u_1,u_2)$ to obtain some approximation results.

Let us first recall that $\psi(x,y,z)$ vanishes only on the image of the set $C = (\frac{1}{2},\frac{1}{2},z)$, $z \in \mathbb{R}$ in $\Gamma\backslash N$, where $(\frac{1}{2},\frac{1}{2},z) \in N$. Furthermore, by the definition of ψ and the classical fact that the theta function $\theta[{}^0_0](\zeta,i)$, $\zeta = x+iy$, has a simple zero at $(\frac{1}{2},\frac{1}{2}) \in \mathbb{C}$, we have easily that there is a neighborhood of C such that

$$\psi(x,y,z) = ((x - \frac{1}{2}) + i(y - \frac{1}{2}))F(x,y,z)$$

where $F(x,y,z)$ is a C^∞ function in C and $F(x,y,z) \neq 0$.

Let us again let $C_1^k = C^k(\Gamma\backslash N) \cap H_1$. Then if $F \in C_1^k$, then F/ψ is clearly a function defined on $\mathbf{T}^2$ and is of class C^r at all points except $(\frac{1}{2},\frac{1}{2})$. In a neighborhood of the point $(\frac{1}{2},\frac{1}{2})$ in $\mathbf{T}^2$ the function of F/ψ can be written as

$$\frac{1}{(x-\frac{1}{2})+i(y-\frac{1}{2})} \times f(x,y)$$

where $f(x,y)$ is a C^r function. Hence $F/\psi \in L^1(\mathbf{T}^2)$. Hence we can apply results from abelian Fourier analysis to study the function F/ψ. We should also note that there does not appear to be any natural condition on F which allows us to infer facts about the singularity of F/ψ at $(\frac{1}{2},\frac{1}{2})$ other than we will mention below. We will see later some examples that illustrate this difficulty.

Now let $F \in C_1^0(\Gamma\backslash N)$ we will now see how the Fourier analysis on $\mathbf{T}^2$ can be used to gain some ability to approximate the function F. Let us begin by recalling the following basic definition: A family of continuous functions $\{K_\epsilon\}_{\epsilon\neq 0}$, on $\mathbf{T}^2$ having absolutely convergent Fourier series is called an approximate identity if it satisfies the following two properties:

(1) $K_\epsilon * p \to p$ uniformly as $\epsilon \to 0$ for every trigonometric polynomial p in $\mathbf{T}^2$, where $*$ denotes convolution.

(2) $\|K_\epsilon\|_1$, the L^1 norm on $\mathbf{T}^2$, is uniformly bounded for all $\epsilon \neq 0$.

Now using Theorem 2.11 on page 253 of [S] we can immediately conclude the following result.

<u>Theorem C.2</u>: For $F \in C_1^0(\Gamma\backslash N)$ and $\{K_\epsilon\}_{\epsilon\neq 0}$ an approximate identity on $\mathbf{T}^2$ the family of functions

$$(K_\epsilon * F/\psi)\psi$$

(a) converges to F in the L^1 norm on $\Gamma\backslash N$ as $\epsilon \to 0$,

(b) converges pointwise to F except perhaps at the zeros of ψ.

In [Z], Zygmund discusses the Fejer kernel on $\mathbf{T}^2$ (see page 303). Let $\{K_n\}$ denote this Fejer kernel. Now let us denote by $C_1^r(a,b)$ the algebra of all functions in $C_1^r(\Gamma\backslash N)$ that vanish on the image of (a,b,z), $z \in \mathbf{R}$ in $\Gamma\backslash N$. It is easy to verify that for $F \in C_1^1(\frac{1}{2},\frac{1}{2})$, the function F/ψ is bounded. Using Theorem 1.20 of [Z] page 304, it is then easy to verify the following result.

<u>Theorem C.3</u>: Let $F \in C_1^1(\frac{1}{2},\frac{1}{2})$ and $\{K_n\}_{n\in\mathbf{Z}^+}$ the Fejer kernel on $\mathbf{T}^2$. Then the sequence

$$(K_n * F/\psi)\psi$$

a) Pointwise converges to F as $n \to \infty$ at all points other than the zeros of ψ.

b) Converges uniformly to F as $n \to \infty$ on any closed set not containing a zero of ψ.

It should be remarked that we could obtain an analogous result to Theorem C.3 for the algebra $C_1^1(a,b)$ by replacing ψ by $R(a',b',0)\psi$, where $(\frac{1}{2},\frac{1}{2},z)(a',b',0) = (a,b,z')$.

Now that we see that Hilbert basis $\psi\chi(u_1,u_2)$, $u_1,u_2 \in \mathbb{Z}$, leads to some sort of analysis on $\Gamma\backslash N$ one is led back to examine how serious a defect is caused by the fact that this basis is not orthonormal. The next few results will show that if we are willing to work with an invariant measure class on $\Gamma\backslash N$ under the action of N, instead of an invariant measure, we can indeed choose a measure μ in the invariant measure class and an orthonormal basis of $L^2(\Gamma\backslash N,\mu)$ may be selected, that is intimately related to the functions $\psi\chi(u_1,u_2)$.

Since $\psi \in H_1$, it is a direct computation to verify that $1/\bar{\psi}$ is an L^1 function $\Gamma\backslash N$ which satisfies the functional equations characterizing functions on H_1.

We have already discussed in Chapter II that $|\psi| \in H_0$ and hence may be viewed as a C^∞ function on $\mathbf{T}^2$. Define a new measure μ on $\Gamma\backslash N$ by the formula

$$\int_{\Gamma\backslash N} F\, d\mu = \int_{\Gamma\backslash N} F|\psi|^2\, dx\, dy\, dz \,.$$

Now by Lemma C.1 the total μ-measure of $\Gamma\backslash N$ is $\sqrt{2}/2$. Since the sets of measure zero of μ are exactly the same as that of Haar measure on $\Gamma\backslash N$ it follows that μ is in the same measure class as Haar measure. Note, too, that the translates of μ under N are equivalent to μ.

We will now compare the spaces $L^2(\Gamma\backslash N),\mu)$ and $L^2(\Gamma\backslash N)$. Let F be a measurable function on $\Gamma\backslash N$ (since μ and Haar measure are equivalent measures this is well defined). Then clearly $F \in L^2(\Gamma\backslash N,\mu)$ if and only if $F|\psi| \in L^2(\Gamma\backslash N)$. It is easily verified that the mapping

$$\varphi: F \to F|\psi|$$

is a linear isomorphism of $L^2(\Gamma\backslash N,\mu)$ onto $L^2(\Gamma\backslash N)$ and it is easily verified that φ is a unitary operator. These observations are summarized in Theorem C.4 below.

Theorem C.4: Let $d\mu = |\psi|^2 \, dx \, dy \, dx$ define a new measure on $\Gamma\backslash N$ and let $L^2(\Gamma\backslash N,\mu)$ be the corresponding Hilbert space of square summable functions. Then μ is equivalent to Haar measure on $\Gamma\backslash N$ and the mapping

$$\varphi: F \to F|\psi|$$

defines a unitary operator of $L^2(\Gamma\backslash N,\mu)$ onto $L^2(\Gamma\backslash N)$.

Define $H_n(\mu)$ to be the space of all functions $F \in L^2(\Gamma\backslash N,\mu)$ such that $F|\psi| \in H_n$. Clearly

$$L^2(\Gamma\backslash N,\mu) = \oplus \ \Sigma_{n\in\mathbb{Z}} H_n(\mu)$$

and the mapping $\varphi: F \to F|\psi|$ induces a unitary operator of $H_n(\mu)$ onto H_n. For the rest of the discussion of this matter, we will restrict our attention to the case where $n = 1$ as we stated was our plan in the introduction to this section. Our plan is to prove that

$$\bar{\psi}^{-1} \chi(u_1,u_2), \quad u_1,u_2 \in \mathbb{Z} ,$$

is an orthonormal basis of $H_1(\mu)$ and see some of the immediate implications of this fact.

Lemma C.5: The function $\bar{\psi}^{-1} \in H_1(\mu)$, where the bar denotes complex conjugate, and has $L^2(\Gamma\backslash N,\mu)$ norm $\|\bar{\psi}^{-1}\|_\mu = 1$.

Proof: Clearly

$$\|\bar{\psi}^{-1}\|_\mu = \int_{\Gamma\backslash N} \frac{1}{|\bar{\psi}|^2} \, d\mu = \int_{\Gamma\backslash N} 1 \, dx \, dy \, dz = 1 .$$

Lemma C.6: Let $p \in L^2(T^2)$. Then the mapping

$$\xi : p \to p/\bar{\psi}$$

defines an isometry of $L^2(T^2)$ onto $H_1(\mu)$.

Proof: For $p,q \in L^2(T^2)$ we have, if $(\ ,\ )_\mu$ denotes the inner product in $H_1(\mu)$,

$$(p/\bar{\psi}, q/\bar{\psi})_\mu = \int_{\Gamma\backslash N} \frac{p\bar{q}}{|\bar{\psi}|^2}\, d\mu = \int_{\pi^2} p\bar{q}\, dx\, dy$$

and we have proven our assertion.

These lemmas prove our previous assertion and may be summarized in the following statement.

Theorem C.6: The functions $e^{2\pi i(u_1 x + u_2 y)} \frac{1}{\bar{\psi}}$, $u_1, u_2 \in \mathbb{Z}$, define an orthonormal basis for $H_1(\mu)$.

Proof: Take $F \in H_1(\mu)$ and assume $(\bar{\psi}^{-1}\chi(u_1,u_2),F)_\mu = 0$ all $u_1,u_2 \in \mathbb{Z}$. Then

$$0 = \int_{\Gamma\backslash N} \chi(u_1,u_2)p\bar{\psi}^{-1}\, \bar{F}\, d\mu = \int_{\Gamma\backslash N} \chi(u_1,u_2)\bar{\psi}^{-1}\, \bar{F}\, \bar{\psi}\psi\, dx\, dy\, dz$$

$$= \int_{\Gamma\backslash N} \chi(u_1,u_2)\psi\, \bar{F}\, dx\, dy\, dz$$

$$= (\chi(u_1,u_2)\psi, F)$$

where the last inner product is in $L^2(\Gamma\backslash N)$ or H_1. Since $\chi(u_1,u_2)\psi$ is a Hilbert space basis of H_1, we have $F = 0$ a.e. with respect to $dx\, dy\, dz$ and so with respect to μ, since $dx\, dy\, dz$ and μ are equivalent measures.

Let $\xi_{mn} = e^{2\pi i(mx+ny)} \frac{1}{\bar{\psi}}$, $m,n \in \mathbb{Z}$. The ξ_{mn} is an orthonormal basis of $H_1(\mu)$.Then for $F \in H_1(\mu)$ define

$$\hat{F}(m,n) = \int_{\Gamma\backslash N} F\, \bar{\xi}_{mn}\, d\mu$$

and call

$$\sum_{m,n\in\mathbf{Z}} \hat{F}(m,n)\, \xi_{mn}$$

the Fourier series of F in $H_1(\mu)$. The following statement requires no proof.

Lemma C.7: For $F \in H_1$, $F\bar{\psi} \in L^2(\mathbf{T}_2)$ and the Fourier coefficient $\hat{F}(m,n)_{\mathbf{T}^2}$ is given by

$$\hat{F}(m,n)_{\mathbf{T}^2} = \int_{\mathbf{T}^2} F\,\bar{\psi}\, e^{-2\pi i(mx+ny)}\, dx\, dy$$

Lemma C.7 has the important implication that if

$$\bar{\psi} F \sim \Sigma\, a_{mn}\, e^{2\pi i(mx+ny)}$$

is the Fourier series of $\bar{\psi} F$ on $\mathbf{T}^2$, then

$$F \sim \Sigma\, a_{mn}\, \xi_{mn}$$

is the Fourier series of F in $H_1(\mu)$.

Theorem C.8: If $F \in C_1^r(\Gamma\backslash N)$, $r \geq 2$ then the Fourier series

$$F \sim \Sigma_{m,n\in\mathbf{Z}}\, \hat{F}(m,n)\, \xi_{mn}$$

converges absolutely to F except perhaps at the points where ψ vanishes.

Proof: Since $F\bar{\psi} \in C^r(\mathbf{T}^2)$, $r \geq 2$, by the classical results in abelian Fourier analysis the Fourier series of $F\bar{\psi}$ converges absolutely to $F\bar{\psi}$. Hence

$$F = \bar{\psi}^{-1}\, \Sigma\, \hat{F}(m,n)\, e^{2\pi i(mx+ny)} = \Sigma\, \hat{F}(m,n)\, \xi_{mn}$$

except at the zeros of the function ψ and the convergence is absolute.

Since $\bar{\psi}^{-1}$ is bounded on any closed set not containing the zeros of $\bar{\psi}$, we get as a corollary of the preceeding theorem the following result.

Corollary C.9: If $F \in C_1^r(\Gamma \backslash N)$, $r \geq 2$, then the Fourier series $\Sigma \hat{F}(m,n)\, \xi_{mn}$ of F converges uniformly to F on closed set not containing a zero of ψ.

As an example we shall compute the Fourier series in $H_1(\mu)$ of the function ψ. As mentioned above this is equivalent to determining the Fourier series of the function $\psi\bar{\psi}$. Now

$$(\psi\bar{\psi})^{\wedge}(m,n)_{T^2} = \int_{T^2} \psi\bar{\psi}\, e^{-2\pi i(mx+ny)}\, dx\, dy$$

$$= (\psi\chi(0,0),\ \psi\chi(m,n))$$

where the product is in H_1. By Lemma C.1 this gives

$$(\psi\bar{\psi})^{\wedge}(m,n)_{T^2} = \sqrt{2/2}\ e^{-(\pi/2)(m^2+n^2)} e^{\pi imn} .$$

Hence on T^2, $\psi\bar{\psi}$, is given by the absolutely convergent Fourier series

$$\psi\bar{\psi}(x,y) = \sqrt{2/2}\ \sum_{n,m\in\mathbb{Z}} e^{-(\pi/2)(m^2+n^2)} e^{\pi imn} e^{2\pi i(mx+ny)}$$

$$= \sqrt{2/2}\ \sum_{k=0}^{\infty} e^{-(\pi/2)k} \sum_{m^2+n^2=k} (-1)^{mn} e^{2\pi i(mx+ny)} .$$

Finally, ψ is given by

$$\psi = \sqrt{2/2}\ \sum_{k=0}^{\infty} e^{-(\pi/2)k} \sum_{m^2+n^2=k} (-1)^{mn} \xi_{mn} .$$

Second Approach: Our second approach is really an effort to refine Theorem II.5. Recall that Theorem II.5 states that for $F \in C_1^r$ we can find $p,q \in C^r(T^2)$ such that

$$F = p\psi + q\psi^* .$$

The problem with this result as we pointed out in Chapter II is that p and q are not uniquely determined by the above condition. We shall discuss in this section, how we can considerably control the function spaces C_1^1 and C_1^2 by refining the above result.

Our plan is to consider the automorphism of $\Gamma\backslash N$ determined by $(x,y,z) \to (-y,x,z-xy)$ discussed in Chapters I and III. Now for $p \in C(T^2) = H_0$ we will denote the affect of the automorphism $(x,y,z) \to (-y,x,z-xy)$ by $T_\sigma\colon p \to p^\sigma$ and similarly for $F \in C_1^0(\Gamma\backslash N)$ we have $T_\sigma\colon F \to F^\sigma$. The functions p^σ and F^σ are then given explicitly by the formulas

$$p^\sigma(x,y) = p(-y,x) \qquad p \in C(T^2)$$

$$F^\sigma(x,y,z) = e^{-2\pi ixy} F(-y,x,z) \qquad F \in C_1^0 .$$

It is easily verified that $T_\sigma^4 = I$, where I is the identity operator, and that T_σ decomposes $C(T^2)$ and C_1^0 each into the direct sum of four eigen vector spaces corresponding to the eigen values ± 1, $\pm i$. Thus

$$C(T^2) = X_1 \oplus X_{-1} \oplus X_i \oplus X_{-i}$$

and

$$C_1^0 = Y_1 \oplus Y_{-1} \oplus Y_i \oplus Y_{-i}$$

where X_α (Y_α) is the eigen vector space with eigen value α, where $\alpha = \pm 1, \pm i$. It is then trivially verified that the projection mappings $q_\alpha\colon C(T^2) \to X_\alpha$ and $q_\alpha^*\colon C_1^0 \to Y_\alpha$ are given by

$$(1)\quad 4q_\alpha(p) = 4\,p_\alpha = p + \alpha p^{-\sigma} + \alpha^2 p^{-\sigma^2} + \alpha^3 p^{-\sigma^3} , \qquad p \in C(T_2)$$

$$(2)\quad 4q_\alpha^*(F) = 4\,F_\alpha = F + \alpha F^{-\sigma} + \alpha^2 F^{-\sigma^2} + \alpha^3 F^{-\sigma^3} , \qquad F \in C_1^0 .$$

We have already verified in Lemma III.3 that $\psi \in Y_1$; i.e., $\psi^\sigma = \psi$. We shall need the explicit projection of ψ^* on the spaces Y_α, $\alpha = \pm 1, \pm i$ and so we shall set out the results of these computations. We first observe that

$$(\psi^*)^\sigma = ie^{2\pi ix}\,\psi^* .$$

It is then a simple matter using formula (2) above to compute ψ_α^*. We have listed the results of this computation explicitly in Lemma C.9 below.

Lemma C.9:

1. $\psi_1^* = \beta_1(x,y)\psi^*$, where

$$\beta_1(x,y) = 1 + ie^{2\pi ix} - e^{2\pi ix}e^{-2\pi iy} - ie^{-2\pi iy}$$

2. $\psi_{-1}^* = \beta_{-1}(x,y)\psi^*$, where

$$\beta_{-1}(x,y) = 1 - ie^{2\pi ix} - e^{2\pi ix}e^{-2\pi iy} + ie^{-2\pi iy}$$

3. $\psi_i^* = \beta_i(x,y)\psi^*$, where

$$\beta_i(x,y) = 1 + e^{2\pi ix} + e^{2\pi ix}e^{-2\pi iy} + e^{-2\pi iy}$$

4. $\psi_{-i}^* = \beta_{-i}(x,y)\psi^*$, where

$$\beta_{-i}(x,y) = 1 - e^{2\pi ix} + e^{2\pi ix}e^{-2\pi iy} - e^{-2\pi iy} .$$

We shall now analyze the zeros and the local behavior or these functions at their zeros. This information will play an important role in our later work.

Case 1: The zeros of $\beta_1(x,y)$.

Setting $\beta_1(x,y) = 0$, we obtain

$$1 + e^{2\pi ix} = e^{-2\pi iy} \cdot i(1 - ie^{2\pi ix}) .$$

This yields

$$|1 + ie^{2\pi ix}| = |1 - ie^{2\pi ix}|$$

and some elementary computations show the above implies

$$e^{2\pi ix} = e^{-2\pi ix} \quad \text{or} \quad x = 0, \tfrac{1}{2} .$$

An elementary substitution then yields that

$$\beta_1(x,y) = 0 \quad \text{if and only if} \quad x = y = \tfrac{1}{2},\ x = y = 0.$$

The function $\beta_1(x,y)$ is real analytic and has globally defined power series expansions at $(0,0)$ and $(\frac{1}{2},\frac{1}{2})$. The linear terms of these expansions are given as follows:

linear term of $\beta_1(x,y)$ at $(0,0)$ is $-2\pi(1+i)(x-iy)$

linear term of $\beta_1(x,y)$ at $(\frac{1}{2},\frac{1}{2})$ is $2\pi(1-i)((x-\frac{1}{2}) + i(y - \frac{1}{2}))$.

Case 2: The zeros of $\beta_{-1}(x,y)$.

Because $\beta_{-1}(x,y) = \overline{\beta_1(-x,-y)}$, where the bar denotes the complex conjugate by Case 1 we have $\beta_1(x,y)$ has zeros exactly at the points $(0,0)$ and $(\frac{1}{2},\frac{1}{2})$ and

linear term of $\beta_{-1}(x,y)$ at $(0,0)$ is $2\pi(1-i)(x+iy)$

linear term of $\beta_{-1}(x,y)$ at $(\frac{1}{2},\frac{1}{2})$ is $-2\pi(1+i)((x - \frac{1}{2})-i(y - \frac{1}{2}))$.

Case 3: The zeros of $\beta_i(x,y)$.

Setting $\beta_i(x,y) = 0$ we find

$$0 = 1 + e^{2\pi ix} + e^{-2\pi iy}(1 + e^{2\pi ix})$$

$$= (1 + e^{2\pi ix})(1 + e^{-2\pi iy}) .$$

Hence $\beta_i(x,y) = 0$ if and only if $x = 1/2$ or $y = 1/2$. Thus the zeros of $\beta_i(x,y)$ occur on the lines $(1/2,y)$ and $(x,1/2)$.

To compute the linear term of $\beta_i(x,y)$ at its zeros we compute

$$\partial/\partial x(\beta_i(x,y)) \quad \text{and} \quad \partial/\partial y(\beta_i(x,y))$$

and find

$$\partial/\partial x(\beta_i) = 2\pi i(e^{2\pi ix} + e^{2\pi i(x-y)})$$

$$\partial/\partial y(\beta_i) = -2\pi i(e^{2\pi i(x-y)} + e^{-2\pi iy})$$

This implies that

linear term of $\beta_i(x,y)$ is 0 for $x = \frac{1}{2} = y$

linear term of $\beta_i(x,y)$ is $-2\pi i(1+e^{-2\pi iy_0})(x-\frac{1}{2})$ at $(\frac{1}{2},y_0) \neq (\frac{1}{2},\frac{1}{2})$.

linear term of $\beta_i(x,y)$ is $2\pi i(1+e^{2\pi ix_0})(y-\frac{1}{2})$ at $(x_0,\frac{1}{2}) \neq (\frac{1}{2},\frac{1}{2})$.

Case 4: The zeros of $\beta_{-i}(x,y)$.

Because $\beta_{-i}(x,y) = \beta_i(x+\frac{1}{2},y+\frac{1}{2})$, we have that $\beta_{-i}(x,y)$ vanishes on the line $(x,0)$ and $(0,y)$ and

linear term of $\beta_{-i}(x,y)$ is 0 at $x = 0 = y$

linear term of $\beta_{-i}(x,y)$ is $-2\pi i(1-e^{-2\pi iy_0})x$ at $(0,y_0) \neq (0,0)$

linear term of $\beta_{-i}(x,y)$ is $2\pi i(1-e^{2\pi ix_0})y$ at $(x_0,0) \neq (0,0)$.

Our aim is ultimately to prove Theorem C.10 as stated below.

Theorem C.10: For $F \in C_1^1(\Gamma\backslash N)$ we can write

$$F = p\psi + c_{-1}\psi_{-1}^* + c_{-i}\psi_{-i}^*$$

where $p \in C^0(T^2)$, $c_{-1}, c_{-i} \in \mathbb{C}$ and this decomposition is unique.

Our plan in proving Theorem C.10 is to study separately the four special cases $F \in Y_\alpha$, $\alpha = \pm 1, \pm i$ and then combine these results to obtain Theorem C.10. Accordingly, the proof of Theorem C.10 is broken into a sequence of lemmas.

Suppose $F \in C_1^1(\Gamma\backslash N) \cap Y_1$. Then we can write

(1) $F = p\psi + q\psi^*$ where $p,q \in C^1(T_2)$

and further we may choose the p and q to satisfy the functional equations.

(2) $p(-y,x) = p(x,y)$ and $q(-y,x) = -i\, e^{-2\pi i x} q(x,y)$.

Because $\psi^\sigma = \psi$ and $\psi^{*\sigma} = i\, e^{2\pi i x}\psi^*$. It is clear that any expression (1) where p,q satisfy (2) will be in $C_1^1(\Gamma\backslash N) \cap Y_1$. The only problem is to prove that every $F \in C_1^1(\Gamma\backslash N) \cap Y_1$ has an expression of the desired sort. Let us now see that this is indeed the case.

Now let

$$F = p_1^\sigma \psi^\sigma + q_1^\sigma \psi^{*\sigma}$$

$$= p_1^\sigma \psi + q_1^\sigma \psi^*$$

then

$$F = q_1^*(F) = (p_1 + p_1^{-\sigma} + p_1^{-\sigma^2} + p_1^{-\sigma^3})\psi + (q_1 + i\, q_1^{-\sigma}\, e^{2\pi i x} - q_1^{-\sigma^2} e^{2\pi i(x-y)} - i q_1^{-\sigma^3} e^{2\pi i y})\psi^*$$

Let

$$p = p_1 + p_1^{-\sigma} + p_1^{-\sigma^2} + p_1^{-\sigma^3}$$

and

$$q = q_1 + i q_1^{-\sigma} e^{2\pi i x} - q_1^{-\sigma^2} e^{2\pi i(x-y)} - i q_1^{-\sigma^3} e^{2\pi i y}.$$

It only remains to prove that q satisfies the functional equation 2.

Now

$$F = p\psi + q\psi^*$$

and

$$F^\sigma = F \quad \text{and} \quad (p\psi)^\sigma = p\psi .$$

Hence $(q\,\psi^*)^\sigma = q\,\psi^*$. But

$$\psi^{*\sigma} = i\, e^{2\pi i x}\psi^*$$

and so

$$q^\sigma = (ie^{2\psi ix})^{-1} q$$

at all points where $\psi^* \neq 0$ and also by continuity at the closure of the set where $\psi^* \neq 0$ which is all of $\Gamma\backslash N$. This proves our assertion.

Let us now consider the implications of the fact that q satisfies the functional equation (2). From this functional equation it follows easily that

$$i \frac{\partial q}{\partial x}(\tfrac{1}{2},\tfrac{1}{2}) = \frac{\partial q}{\partial y}(\tfrac{1}{2},\tfrac{1}{2}) .$$

Hence, since any function in X_1 vanishes at $(\frac{1}{2},\frac{1}{2})$ we may write

$$q(x,y) = \frac{\partial q}{\partial x}(\tfrac{1}{2},\tfrac{1}{2})(x - \tfrac{1}{2} + i(y - \tfrac{1}{2})) + \varepsilon(x,y)$$

where

$$\lim_{(x,y)\to(\frac{1}{2},\frac{1}{2})} \frac{\varepsilon(x,y)}{\sqrt{(x-\frac{1}{2})^2+(y-\frac{1}{2})^2}} = 0 .$$

It follows that

$$\frac{q(x,y)}{(x-\frac{1}{2})+i(y-\frac{1}{2})}$$

is continuous at $(\frac{1}{2},\frac{1}{2})$. From this it follows that

$$\frac{q(x,y)\psi^*}{\psi} \in C^0(T^2)$$

and so we have proven our first lemma.

<u>Lemma C.11</u>: For $F \in C_1^1 \cap Y_1$ we can uniquely write

$$F = p\psi$$

where $p \in C^0(T^2)$.

Now choose $F \in C_1^1(\Gamma\backslash N) \cap Y_{-1}$. Using the same reasoning as above we can prove that we can write

$$F = p_1\psi + q\psi^*$$

where

$$p,q \in C_1^1(T^2)$$

and satisfy the functional equations

$$(3) \qquad p_1(-y,x) = p_1(x,y) \quad \text{and} \quad q(-y,x) = ie^{-2\pi i x} q(x,y).$$

We again verify that $q(\frac{1}{2},\frac{1}{2}) = 0$ and that

$$\frac{\partial q}{\partial y}(\tfrac{1}{2},\tfrac{1}{2}) = -i\,\frac{\partial q}{\partial x}(\tfrac{1}{2},\tfrac{1}{2}) .$$

It follows that

$$q^{\#}(x,y) = q(x,y) - c\beta_{-1}(x,y) \in \mathbb{C}\,(T^2)$$

and $q^{\#}(\frac{1}{2},\frac{1}{2}) = 0$ as well as

$$\frac{\partial}{\partial x}\, q^{\#}(\tfrac{1}{2},\tfrac{1}{2}) = \frac{\partial}{\partial y} q^{\#}(\tfrac{1}{2},\tfrac{1}{2}) = 0.$$

This implies that

$$\frac{(q(x,y) - c\beta_{-1}(x,y))\psi^*}{\psi} \in \mathbb{C}(T^2) .$$

<u>Lemma C.12</u>: For $F \in \mathbb{C}_1^1(\Gamma\backslash N) \cap Y_{-1}$, we can write

$$F = p\psi + c\psi_{-1}^* \qquad c \in \mathbf{C}\ ,\ p \in \mathbb{C}(T^2)$$

and the decomposition is unique.

<u>Proof:</u> The only thing to prove is the uniqueness part of the lemma. However, since $p\psi = c\psi_{-1}^*$ with $p \in \mathbb{C}(T^2)$ implies

$$p = \frac{c\psi_{-1}^*}{\psi}$$

is continuous. But this is impossible since $\frac{\psi_{-1}^*}{\psi}$ cannot be continuous at $(\frac{1}{2},\frac{1}{2})$. This proves our assertion.

Now let $F \in \mathbb{C}_1^1(\Gamma\backslash N) \cap Y_i$ we can write

$$F = p_1\psi + q\psi^* \qquad p,q \in \mathbb{C}^1(T^2)$$

where p_1 and q satisfy the functional equations

$$(4) \quad p(-y,x) = ip(x,y) \quad \text{and} \quad q(-y,x) = e^{-2\pi ix}q(x,y).$$

(The proof of this assertion is again "as before".)

In this one verifies that

$$q(\tfrac{1}{2},\tfrac{1}{2}) = \frac{\partial q}{\partial x}(\tfrac{1}{2},\tfrac{1}{2}) = \frac{\partial q}{\partial y}(\tfrac{1}{2},\tfrac{1}{2}) = 0$$

So that arguing as before, it follows that

$$\frac{q(x,y)\ \psi^*(x,y,z)}{\psi(x,y,z)} \in \mathcal{C}(T^2)$$

Thus we have outlined how to verify

Lemma C.13: For $F \in C_1^1(\Gamma\backslash N) \cap Y_i$ we can uniquely write

$$F = p\psi \qquad\qquad p \in \mathcal{C}(T^2)$$

Finally, if $F \in C_1^1(\Gamma\backslash N) \cap Y_{-i}$ we can write

$$F = p_1\psi + q\psi^* \qquad \text{where } p_2, q \in C^1(T^2)$$

and p_1 and q satisfy the functional equations

$$(5) \quad p(-y,x) = -ip(x,y) \quad \text{and} \quad q(-y,x) = -e^{-2\pi ix}q(x,y).$$

We verify that $q(\frac{1}{2},\frac{1}{2})$ is not necessarily zero. However, since

$$\beta_{-i}(\tfrac{1}{2},\tfrac{1}{2}) \neq 0$$

we can argue similarly to the eigen-value -1 space to conclude the following lemma.

Lemma C.14: For $F \in C_1^1 \cap Y_{-i}$ we can write

$$F = p\psi + c\cdot\psi_{-1}^*$$

where $p \in \mathcal{C}(T^2)$ and $c \in \mathbb{C}$. Further, this decomposition is unique.

We have thus finally come to a proof of Theorem C.10. It is

interesting to see if we cannot improve the smoothness of the function p occurring in Theorem C.10 by improving the smoothness of the function $F \in H_1$. We will now see that this is indeed the case, but only at a price. To make this last remark clear let us now state the theorem whose proof will be the goal of the last part of these notes.

Theorem C.15: Let $F \in C_1^2$ then we can uniquely write

$$F = p\psi + c_1\psi_1^* + c_{-1}\psi_{-1}^* + c_i\psi_i^* + \gamma_i c_{-i}\psi_{-i}^*$$

where $c_\alpha \in \mathbb{C}$, $\alpha = \pm 1, \pm i$, and $p \in C^1(T^2)$ and $\gamma_i(x,y)$ is defined before the statement of Lemma C.21.

Thus to get p smoother we have had to pay the "price" of allowing a larger, although still finite, "defect" space. Of course, this raises the question of - is there a "good" theory for C^∞ functions? This is still an open question as of this writing.

The proof of Theorem C.15 follows in broad outline the proof of Theorem C.10 but, of course, more care is necessary at each step.

First, consider $F \in C_1^2(\Gamma\backslash N) \cap Y_1$ and as in the proof of Theorem C.10, write

$$F = p_1\psi + q\psi^*$$

where $p,q \in C^2(T^2)$ and satisfy the usual functional equations

$$p(-y,x) = p(x,y) \quad \text{and} \quad q(-y,x) = -ie^{-2\pi ix}q(x,y).$$

To prove that q/ψ, $\psi^* \in C^1(T^2)$ it is clearly enough to prove the following lemma.

Lemma C.16: For $q \in C^2(T^2)$ and satisfying the functional equation $q(-y,x) = ie^{-2\pi ix}q(x,y)$, we have

$$\frac{q((x-\frac{1}{2}),(y-\frac{1}{2}))}{(x-\frac{1}{2})+i(y-\frac{1}{2})}$$

is class C^1 in a neighborhood of the point $(\frac{1}{2},\frac{1}{2})$.

Proof: The functional equation for $q(x,y)$ immediately implies the following four relations:

$$q(\tfrac{1}{2},\tfrac{1}{2}) = 0 \ ; \qquad \frac{\partial q}{\partial y}(\tfrac{1}{2},\tfrac{1}{2}) = i\,\frac{\partial q}{\partial x}(\tfrac{1}{2},\tfrac{1}{2})$$

$$\frac{\partial^2 q}{\partial x^2}(\tfrac{1}{2},\tfrac{1}{2}) = i\,\frac{\partial^2 q}{\partial y^2}(\tfrac{1}{2},\tfrac{1}{2}) = 2\pi\,\frac{\partial q}{\partial x}(\tfrac{1}{2},\tfrac{1}{2})$$

$$\frac{\partial^2 q}{\partial x,\partial y}\,(\tfrac{1}{2},\tfrac{1}{2}) = -\pi(1+i)\,\frac{\partial q}{\partial x}\,(\tfrac{1}{2},\tfrac{1}{2}) \ .$$

Let $\tilde{x} = x - \frac{1}{2}$, $\tilde{y} = y - \frac{1}{2}$ and

$$h(x,y) = \frac{q(x,y)}{\tilde{x}+i\tilde{y}} = \frac{(\tilde{x}-i\tilde{y})\;q(x,y)}{\tilde{x}^2+\tilde{y}^2}$$

Then

$$\frac{\partial}{\partial x}\,h(x,y) = \frac{(\tilde{x}^2+\tilde{y}^2)((\tilde{x}-i\tilde{y})(\frac{\partial q}{\partial x}(x,y))+q(x,y))-2\tilde{x}(\tilde{x}-i\tilde{y})q(x,y)}{(\tilde{x}^2+\tilde{y}^2)^2}$$

Using Taylor's formula for $q(x,y)$ and $\frac{\partial}{\partial x}\,q(x,y)$ near $(\frac{1}{2},\frac{1}{2})$ we get recalling $q(\frac{1}{2},\frac{1}{2}) = 0$

$$q(x,y) = \tilde{x}\,\frac{\partial}{\partial x}\,q(\tfrac{1}{2},\tfrac{1}{2})+\tilde{y}\,\frac{\partial}{\partial y}q(\tfrac{1}{2},\tfrac{1}{2})+\frac{\tilde{x}^2}{2}\,\frac{\partial^2 q}{\partial x^2}(w)+\tilde{x}\tilde{y}\,\frac{\partial^2}{\partial x\partial y}\,q(w)+\frac{\tilde{y}^2}{2}\,\frac{\partial^2}{\partial y^2}q(w)$$

and

$$\frac{\partial}{\partial x}q(x,y) = \frac{\partial q}{\partial x}(\tfrac{1}{2},\tfrac{1}{2}) + \tilde{x}\,\frac{\partial^2}{\partial x^2}q(v) + \tilde{y}\,\frac{\partial^2}{\partial x\partial y}\,q(v)$$

where w,v are points on the line joining $(\frac{1}{2},\frac{1}{2})$ to (x,y). Substituting into the expression for $\frac{\partial}{\partial x}\,h(x,y)$ we have that

$$\frac{\partial}{\partial x}h(x,y) = \tilde{x}^4\left(\frac{\partial^2}{\partial x}q(v) - \tfrac{1}{2}\frac{\partial^2}{\partial x^2}q(w)\right)$$

$$+ \tilde{x}^3\tilde{y}\left(\frac{\partial^2}{\partial x\partial y}q(v) - i\frac{\partial^2}{\partial x^2}q(v)\frac{\partial^2}{\partial x\partial y}q(w) + i\frac{\partial^2}{\partial x^2}q(w)\right)$$

$$+ \tilde{x}^2\tilde{y}^2\left(-i\frac{\partial^2}{\partial x\partial y}q(v) + \frac{\partial^2}{\partial x^2}q(v) + \tfrac{1}{2}\frac{\partial^2}{\partial y^2}q(w) + \tfrac{1}{2}\frac{\partial^2}{\partial x^2}q(w)\right.$$

$$\left.- \frac{\partial^2}{\partial y^2}q(w) + 2i\frac{\partial^2}{\partial x\partial y}q(w)\right)$$

$$+ \tilde{x}\tilde{y}^3\left(\frac{\partial^2}{\partial x\partial y}q(v) - i\frac{\partial^2}{\partial x^2}q(v) + \frac{\partial^2}{\partial x\partial y}q(w) + i\frac{\partial^2}{\partial y^2}q(w)\right)$$

$$+ \tilde{y}^4\left(-i\frac{\partial^2}{\partial x\partial y}q(v) + \tfrac{1}{2}\frac{\partial^2}{\partial y^2}q(w)\right).$$

Since $q \in C^2$ and

$$\frac{\tilde{x}^4}{(\tilde{x}^2+\tilde{y}^2)^2}, \quad \frac{\tilde{x}^3\tilde{y}}{(\tilde{x}^2+\tilde{y}^2)^2} \qquad \text{etc.}$$

are bounded it is easy to see that

$$\lim_{(x,y)\to(\frac{1}{2},\frac{1}{2})} \frac{\partial h}{\partial x}(x,y)$$

is given by replacing v and w by $(\frac{1}{2},\frac{1}{2})$ in the above expression for

$$\frac{\partial h}{\partial x}(x,y).$$

It follows from the relations existing between the partial derivatives expressed above that this limit is equal to

$$\pi i \frac{\partial}{\partial x} q(\tfrac{1}{2},\tfrac{1}{2}) .$$

Similarly for

$$\frac{\partial}{\partial y} h(x,y) .$$

Hence we have the following result.

<u>Lemma C.17</u>: For every $F \in C_1^2(\Gamma\backslash N) \cap Y_1$ there exists a unique $p \in C^1(T^2)$ such that $F = p\psi$.

We remark that this result cannot be improved upon. For consider $F = \psi_1^* = \beta_1\psi^*$. Locally, about $(\frac{1}{2},\frac{1}{2})$ we have that β_1 is real analytic and the first three terms of its expansion are given by

$$\begin{aligned}\beta_1(x,y) &= 2\pi(1 - i)((x - \tfrac{1}{2}) + i(y - \tfrac{1}{2}))\\ &+ 2\pi^2(i((x-\tfrac{1}{2}) + i(y-\tfrac{1}{2}))^2\\ &+ ((x-\tfrac{1}{2}) + i(y-\tfrac{1}{2}))((x-\tfrac{1}{2})-i(y-\tfrac{1}{2})))\\ &+ 8\pi^3(((x-\tfrac{1}{2}) + i(y-\tfrac{1}{2}))((x-\tfrac{1}{2})^2\\ &+ (1+2i)(x-\tfrac{1}{2})(y-\tfrac{1}{2}) + (y-\tfrac{1}{2})^2))\\ &+ 3(1-i)(x-\tfrac{1}{2})(y-\tfrac{1}{2})^2\\ &+ \dots\end{aligned}$$

If

$$\frac{\beta_1(x,y)}{(x-\frac{1}{2})+i(y-\frac{1}{2})} \in C^2 ,$$

then clearly

$$\frac{(x-\frac{1}{2})(y-\frac{1}{2})^2}{(x-\frac{1}{2})+i(y-\frac{1}{2})}$$

is class C^2 which is false and so we have the required contradiction.

Consider $F \in C_1^2(\Gamma\backslash N) \cap Y_{-1}$. Then we may choose $p_1, q \in C^2(T^2)$ such that

$$F = p_1\psi + q\psi^*$$

and

$$p_1(-y,x) = -p_1(x,y) \quad \text{and} \quad q(-y,x) = ie^{-2\pi ix}q(x,y) .$$

We have to show that if $\frac{\partial q}{\partial x}(\frac{1}{2},\frac{1}{2}) = 0$, then $\frac{q}{\psi}\,\psi^* \in C^1(T^2)$. This is equivalent to the following lemma.

Lemma C.18: For $q \in C^2(T^2)$ satisfying the function equation

$$q(-y,x) = -ie^{-2\pi ix}q(x,y) \quad \text{and} \quad \frac{\partial q}{\partial x}(0,0) = 0$$

then

$$\frac{q}{x+iy} \in C^1$$

in some neighborhood of $(0,0)$.

Proof: The functional equation and $\frac{\partial q}{\partial x}(0,0) = 0$ implies that all the first and second partial derivatives at $(0,0)$ vanish. This fact easily implies the lemma.

It is now fairly easy to show that the following is true.

Lemma C.19: For $F \in C_1^2(\Gamma\backslash N) \cap Y_{-1}$ there exists a unique $p \in C^1(T^2)$ and a $c \in \mathbb{C}$ such that

$$F = p\psi + c\psi_{-1}^*$$

For $F \in C_1^2(\Gamma\backslash N) \cap Y_{-i}$ there exists a unique $p \in C^1(T^2)$ and a $c \in \mathbb{C}$ such that

$$F = p\psi + c\psi_{-i}^* .$$

In the case where $F \in C_1^2(\Gamma\backslash N) \cap Y_i$ one must work a little harder to obtain the desired results. Again begin by writing

$$F = p_1\psi + q\psi^*$$

where p_1 and q are chosen so as to satisfy the function equations

$$p(-y,x) = ip(x,y) \quad \text{and} \quad q(-y,x) = e^{-2\pi ix}q(x,y)$$

and

$$p,q \in C^2(T^2) .$$

The following lemma is then easily verified.

Lemma C.20: For $q \in C^2(T^2)$ satisfying the functional equation

$$q(-y,x) = -e^{-2\pi i x} q(x,y)$$

if we assume

$$\frac{\partial^2 q}{\partial x^2}(0,0) = 0 \quad \text{and} \quad \frac{\partial^2 q}{\partial x \partial y}(0,0) = 0$$

then

$$\frac{q(x,y)}{x+iy} \text{ is } C^1 \text{ in a neighborhood of } (0,0).$$

Note that $\beta_{-1} \in C^\infty(T^2)$ and β_{-1} vanishes at $(0,0)$ and the linear term of β_{-1} behaves like xy. Rotating β_{-1} by 45°, we get the function

$$\beta_{-1}(x+y,-x+y) .$$

Set $\gamma_{-i} = -e^{-2\pi i x} \beta_{-i}(x+y\ ,-x+y)$. Then γ_{-i} vanishes at $(0,0)$ and its linear term at $(0,0)$ behaves like $x^2 - y^2$. It follows that for any $q \in C^2(T^2)$ satisfying

$$q(-y,x) = -e^{-2\pi i x} q(x,y).$$

There exists constants $c,d \in \mathbb{C}$ such that

$$q - c\,\beta_{-i} - d\,\gamma_{-i}$$

satisfy the hypothesis of Lemma C.20. Hence setting

$$\gamma_i = \gamma_{-i}(x + \tfrac{1}{2}, y + \tfrac{1}{2})$$

we have the following lemma.

Lemma C.21: For $F \in \mathcal{C}_1^2 \cap Y_i$ we can write F uniquely as

$$F = p\psi + c\psi_i^* + d\gamma_i \psi_{-i}^*$$

where $p \in \mathcal{C}^1(T^2)$, $c,d \in \mathbb{C}$ and γ_i is as defined above.

The preceeding lemmas combine to prove Theorem C.15.

REFERENCES

[B] R. Bellman. A Brief Introduction to Theta Functions, Holt, Rinehardt and Winston, New York, 1961.

[S] E. Stein and G. Weiss. Fourier Analysis on Euclidean Space, Princeton Mathematical Series, Princeton Press, Princeton (1971).

[Z] A. Zygmund. Trigonometric Series, Volume II, second edition, Cambridge University Press.

Vol. 277: Séminaire Banach. Edité par C. Houzel. VII, 229 pages. 1972. DM 20,–

Vol. 278: H. Jacquet, Automorphic Forms on GL(2). Part II. XIII, 142 pages. 1972. DM 16,–

Vol. 279: R. Bott, S. Gitler and I. M. James, Lectures on Algebraic and Differential Topology. V, 174 pages. 1972. DM 18,–

Vol. 280: Conference on the Theory of Ordinary and Partial Differential Equations. Edited by W. N. Everitt and B. D. Sleeman. XV, 367 pages. 1972. DM 26,–

Vol. 281: Coherence in Categories. Edited by S. Mac Lane. VII, 235 pages. 1972. DM 20,–

Vol. 282: W. Klingenberg und P. Flaschel, Riemannsche Hilbertmannigfaltigkeiten. Periodische Geodätische. VII, 211 Seiten. 1972. DM 20,–

Vol. 283: L. Illusie, Complexe Cotangent et Déformations II. VII, 304 pages. 1972. DM 24,–

Vol. 284: P. A. Meyer, Martingales and Stochastic Integrals I. VI, 89 pages. 1972. DM 16,–

Vol. 285: P. de la Harpe, Classical Banach-Lie Algebras and Banach-Lie Groups of Operators in Hilbert Space. III, 160 pages. 1972. DM 16,–

Vol. 286: S. Murakami, On Automorphisms of Siegel Domains. V, 95 pages. 1972. DM 16,–

Vol. 287: Hyperfunctions and Pseudo-Differential Equations. Edited by H. Komatsu. VII, 529 pages. 1973. DM 36,–

Vol. 288: Groupes de Monodromie en Géométrie Algébrique. (SGA 7 I). Dirigé par A. Grothendieck. IX, 523 pages. 1972. DM 50,–

Vol. 289: B. Fuglede, Finely Harmonic Functions. III, 188. 1972. DM 18,–

Vol. 290: D. B. Zagier, Equivariant Pontrjagin Classes and Applications to Orbit Spaces. IX, 130 pages. 1972. DM 16,–

Vol. 291: P. Orlik, Seifert Manifolds. VIII, 155 pages. 1972. DM 16,–

Vol. 292: W. D. Wallis, A. P. Street and J. S. Wallis, Combinatorics: Room Squares, Sum-Free Sets, Hadamard Matrices. V, 508 pages. 1972. DM 50,–

Vol. 293: R. A. DeVore, The Approximation of Continuous Functions by Positive Linear Operators. VIII, 289 pages. 1972. DM 24,–

Vol. 294: Stability of Stochastic Dynamical Systems. Edited by R. F. Curtain. IX, 332 pages. 1972. DM 26,–

Vol. 295: C. Dellacherie, Ensembles Analytiques, Capacités, Mesures de Hausdorff. XII, 123 pages. 1972. DM 16,–

Vol. 296: Probability and Information Theory II. Edited by M. Behara, K. Krickeberg and J. Wolfowitz. V, 223 pages. 1973. DM 20,–

Vol. 297: J. Garnett, Analytic Capacity and Measure. IV, 138 pages. 1972. DM 16,–

Vol. 298: Proceedings of the Second Conference on Compact Transformation Groups. Part 1. XIII, 453 pages. 1972. DM 32,–

Vol. 299: Proceedings of the Second Conference on Compact Transformation Groups. Part 2. XIV, 327 pages. 1972. DM 26,–

Vol. 300: P. Eymard, Moyennes Invariantes et Représentations Unitaires. II. 113 pages. 1972. DM 16,–

Vol. 301: F. Pittnauer, Vorlesungen über asymptotische Reihen. VI, 186 Seiten. 1972. DM 18,–

Vol. 302: M. Demazure, Lectures on p-Divisible Groups. V, 98 pages. 1972. DM 16,–

Vol. 303: Graph Theory and Applications. Edited by Y. Alavi, D. R. Lick and A. T. White. IX, 329 pages. 1972. DM 26,–

Vol. 304: A. K. Bousfield and D. M. Kan, Homotopy Limits, Completions and Localizations. V, 348 pages. 1972. DM 26,–

Vol. 305: Théorie des Topos et Cohomologie Etale des Schémas. Tome 3. (SGA 4). Dirigé par M. Artin, A. Grothendieck et J. L. Verdier. VI, 640 pages. 1973. DM 50,–

Vol. 306: H. Luckhardt, Extensional Gödel Functional Interpretation. VI, 161 pages. 1973. DM 18,–

Vol. 307: J. L. Bretagnolle, S. D. Chatterji et P.-A. Meyer, Ecole d'été de Probabilités: Processus Stochastiques. VI, 198 pages. 1973. DM 20,–

Vol. 308: D. Knutson, λ-Rings and the Representation Theory of the Symmetric Group. IV, 203 pages. 1973. DM 20,–

Vol. 309: D. H. Sattinger, Topics in Stability and Bifurcation Theory. VI, 190 pages. 1973. DM 18,–

Vol. 310: B. Iversen, Generic Local Structure of the Morphisms in Commutative Algebra. IV, 108 pages. 1973. DM 16,–

Vol. 311: Conference on Commutative Algebra. Edited by J. W. Brewer and E. A. Rutter. VII, 251 pages. 1973. DM 22,–

Vol. 312: Symposium on Ordinary Differential Equations. Edited by W. A. Harris, Jr. and Y. Sibuya. VIII, 204 pages. 1973. DM 22,–

Vol. 313: K. Jörgens and J. Weidmann, Spectral Properties of Hamiltonian Operators. III, 140 pages. 1973. DM 16,–

Vol. 314: M. Deuring, Lectures on the Theory of Algebraic Functions of One Variable. VI, 151 pages. 1973. DM 16,–

Vol. 315: K. Bichteler, Integration Theory (with Special Attention to Vector Measures). VI, 357 pages. 1973. DM 26,–

Vol. 316: Symposium on Non-Well-Posed Problems and Logarithmic Convexity. Edited by R. J. Knops. V, 176 pages. 1973. DM 18,–

Vol. 317: Séminaire Bourbaki – vol. 1971/72. Exposés 400–417. IV, 361 pages. 1973. DM 26,–

Vol. 318: Recent Advances in Topological Dynamics. Edited by A. Beck, VIII, 285 pages. 1973. DM 24,–

Vol. 319: Conference on Group Theory. Edited by R. W. Gatterdam and K. W. Weston. V, 188 pages. 1973. DM 18,–

Vol. 320: Modular Functions of One Variable I. Edited by W. Kuyk. V, 195 pages. 1973. DM 18,–

Vol. 321: Séminaire de Probabilités VII. Edité par P. A. Meyer. VI, 322 pages. 1973. DM 26,–

Vol. 322: Nonlinear Problems in the Physical Sciences and Biology. Edited by I. Stakgold, D. D. Joseph and D. H. Sattinger. VIII, 357 pages. 1973. DM 26,–

Vol. 323: J. L. Lions, Perturbations Singulières dans les Problèmes aux Limites et en Contrôle Optimal. XII, 645 pages. 1973. DM 42,–

Vol. 324: K. Kreith, Oscillation Theory. VI, 109 pages. 1973. DM 16,–

Vol. 325: Ch.-Ch. Chou, La Transformation de Fourier Complexe et L'Equation de Convolution. IX, 137 pages. 1973. DM 16,–

Vol. 326: A. Robert, Elliptic Curves. VIII, 264 pages. 1973. DM 22,–

Vol. 327: E. Matlis, 1-Dimensional Cohen-Macaulay Rings. XII, 157 pages. 1973. DM 18,–

Vol. 328: J. R. Büchi and D. Siefkes, The Monadic Second Order Theory of All Countable Ordinals. VI, 217 pages. 1973. DM 20,–

Vol. 329: W. Trebels, Multipliers for (C, α)-Bounded Fourier Expansions in Banach Spaces and Approximation Theory. VII, 103 pages. 1973. DM 16,–

Vol. 330: Proceedings of the Second Japan-USSR Symposium on Probability Theory. Edited by G. Maruyama and Yu. V. Prokhorov. VI, 550 pages. 1973. DM 36,–

Vol. 331: Summer School on Topological Vector Spaces. Edited by L. Waelbroeck. VI, 226 pages. 1973. DM 20,–

Vol. 332: Séminaire Pierre Lelong (Analyse) Année 1971-1972. V, 131 pages. 1973. DM 16,–

Vol. 333: Numerische, insbesondere approximationstheoretische Behandlung von Funktionalgleichungen. Herausgegeben von R. Ansorge und W. Törnig. VI, 296 Seiten. 1973. DM 24,–

Vol. 334: F. Schweiger, The Metrical Theory of Jacobi-Perron Algorithm. V, 111 pages. 1973. DM 16,–

Vol. 335: H. Huck, R. Roitzsch, U. Simon, W. Vortisch, R. Walden, B. Wegner und W. Wendland, Beweismethoden der Differentialgeometrie im Großen. IX, 159 Seiten. 1973. DM 18,–

Vol. 336: L'Analyse Harmonique dans le Domaine Complexe. Edité par E. J. Akutowicz. VIII, 169 pages. 1973. DM 18,–

Vol. 337: Cambridge Summer School in Mathematical Logic. Edited by A. R. D. Mathias and H. Rogers. IX, 660 pages. 1973. DM 42,–

Vol: 338: J. Lindenstrauss and L. Tzafriri, Classical Banach Spaces. IX, 243 pages. 1973. DM 22,–

Vol. 339: G. Kempf, F. Knudsen, D. Mumford and B. Saint-Donat, Toroidal Embeddings I. VIII, 209 pages. 1973. DM 20,–

Vol. 340: Groupes de Monodromie en Géométrie Algébrique. (SGA 7 II). Par P. Deligne et N. Katz. X, 438 pages. 1973. DM 40,–

Vol. 341: Algebraic K-Theory I, Higher K-Theories. Edited by H. Bass. XV, 335 pages. 1973. DM 26,–

Vol. 342: Algebraic K-Theory II, "Classical" Algebraic K-Theory, and Connections with Arithmetic. Edited by H. Bass. XV, 527 pages. 1973. DM 36,–

Vol. 343: Algebraic K-Theory III, Hermitian K-Theory and Geometric Applications. Edited by H. Bass. XV, 572 pages. 1973. DM 38,–

Vol. 344: A. S. Troelstra (Editor), Metamathematical Investigation of Intuitionistic Arithmetic and Analysis. XVII, 485 pages. 1973. DM 34,–

Vol. 345: Proceedings of a Conference on Operator Theory. Edited by P. A. Fillmore. VI, 228 pages. 1973. DM 20,–

Vol. 346: Fučík et al., Spectral Analysis of Nonlinear Operators. II, 287 pages. 1973. DM 26,–

Vol. 347: J. M. Boardman and R. M. Vogt, Homotopy Invariant Algebraic Structures on Topological Spaces. X, 257 pages. 1973. DM 22,–

Vol. 348: A. M. Mathai and R. K. Saxena, Generalized Hypergeometric Functions with Applications in Statistics and Physical Sciences. VII, 314 pages. 1973. DM 26,–

Vol. 349: Modular Functions of One Variable II. Edited by W. Kuyk and P. Deligne. V, 598 pages. 1973. DM 38,–

Vol. 350: Modular Functions of One Variable III. Edited by W. Kuyk and J.-P. Serre. V, 350 pages. 1973. DM 26,–

Vol. 351: H. Tachikawa, Quasi-Frobenius Rings and Generalizations. XI, 172 pages. 1973. DM 18,–

Vol. 352: J. D. Fay, Theta Functions on Riemann Surfaces. V, 137 pages. 1973. DM 16,–

Vol. 353: Proceedings of the Conference on Orders, Group Rings and Related Topics. Organized by J. S. Hsia, M. L. Madan and T. G. Ralley. X, 224 pages. 1973. DM 20,–

Vol. 354: K. J. Devlin, Aspects of Constructibility. XII, 240 pages. 1973. DM 22,–

Vol. 355: M. Sion, A Theory of Semigroup Valued Measures. V, 140 pages. 1973. DM 16,–

Vol. 356: W. L. J. van der Kallen, Infinitesimally Central-Extensions of Chevalley Groups. VII, 147 pages. 1973. DM 16,–

Vol. 357: W. Borho, P. Gabriel und R. Rentschler, Primideale in Einhüllenden auflösbarer Lie-Algebren. V, 182 Seiten. 1973. DM 18,–

Vol. 358: F. L. Williams, Tensor Products of Principal Series Representations. VI, 132 pages. 1973. DM 16,–

Vol. 359: U. Stammbach, Homology in Group Theory. VIII, 183 pages. 1973. DM 18,–

Vol. 360: W. J. Padgett and R. L. Taylor, Laws of Large Numbers for Normed Linear Spaces and Certain Fréchet Spaces. VI, 111 pages. 1973. DM 16,–

Vol. 361: J. W. Schutz, Foundations of Special Relativity: Kinematic Axioms for Minkowski Space Time. XX, 314 pages. 1973. DM 26,–

Vol. 362: Proceedings of the Conference on Numerical Solution of Ordinary Differential Equations. Edited by D. Bettis. VIII, 490 pages. 1974. DM 34,–

Vol. 363: Conference on the Numerical Solution of Differential Equations. Edited by G. A. Watson. IX, 221 pages. 1974. DM 20,–

Vol. 364: Proceedings on Infinite Dimensional Holomorphy. Edited by T. L. Hayden and T. J. Suffridge. VII, 212 pages. 1974. DM 20,–

Vol. 365: R. P. Gilbert, Constructive Methods for Elliptic Equations. VII, 397 pages. 1974. DM 26,–

Vol. 366: R. Steinberg, Conjugacy Classes in Algebraic Groups (Notes by V. V. Deodhar). VI, 159 pages. 1974. DM 18,–

Vol. 367: K. Langmann und W. Lütkebohmert, Cousinverteilungen und Fortsetzungssätze. VI, 151 Seiten. 1974. DM 16,–

Vol. 368: R. J. Milgram, Unstable Homotopy from the Stable Point of View. V, 109 pages. 1974. DM 16,–

Vol. 369: Victoria Symposium on Nonstandard Analysis. Edited by A. Hurd and P. Loeb. XVIII, 339 pages. 1974. DM 26,–

Vol. 370: B. Mazur and W. Messing, Universal Extensions and One Dimensional Crystalline Cohomology. VII, 134 pages. 1974. DM 16,–

Vol. 371: V. Poenaru, Analyse Différentielle. V, 228 pages. 1974. DM 20,–

Vol. 372: Proceedings of the Second International Conference on the Theory of Groups 1973. Edited by M. F. Newman. VII, 740 pages. 1974. DM 48,–

Vol. 373: A. E. R. Woodcock and T. Poston, A Geometrical Study of the Elementary Catastrophes. V, 257 pages. 1974. DM 22,–

Vol. 374: S. Yamamuro, Differential Calculus in Topologica Linear Spaces. IV, 179 pages. 1974. DM 18,–

Vol. 375: Topology Conference 1973. Edited by R. F. Dickman Jr. and P. Fletcher. X, 283 pages. 1974. DM 24,–

Vol. 376: D. B. Osteyee and I. J. Good, Information, Weight of Evidence, the Singularity between Probability Measures and Signal Detection. XI, 156 pages. 1974. DM 16,–

Vol. 377: A. M. Fink, Almost Periodic Differential Equations VIII, 336 pages. 1974. DM 26,–

Vol. 378: TOPO 72 – General Topology and its Applications Proceedings 1972. Edited by R. Alò, R. W. Heath and J. Nagata. XIV, 651 pages. 1974. DM 50,–

Vol. 379: A. Badrikian et S. Chevet, Mesures Cylindriques Espaces de Wiener et Fonctions Aléatoires Gaussiennes. X 383 pages. 1974. DM 32,–

Vol. 380: M. Petrich, Rings- and Semigroups. VIII, 182 pages 1974. DM 18,–

Vol. 381: Séminaire de Probabilités VIII. Edité par P. A. Meyer IX, 354 pages. 1974. DM 32,–

Vol. 382: J. H. van Lint, Combinatorial Theory Seminar Eindhoven University of Technology. VI, 131 pages. 1974. DM 18,–

Vol. 383: Séminaire Bourbaki – vol. 1972/73. Exposés 418-435 IV, 334 pages. 1974. DM 30,–

Vol. 384: Functional Analysis and Applications, Proceedings 1972. Edited by L. Nachbin. V, 270 pages. 1974. DM 22,–

Vol. 385: J. Douglas Jr. and T. Dupont, Collocation Methods fo Parabolic Equations in a Single Space Variable (Based on C Piecewise-Polynomial Spaces). V, 147 pages. 1974. DM 16,–

Vol. 386: J. Tits, Buildings of Spherical Type and Finite BN Pairs. IX, 299 pages. 1974. DM 24,–

Vol. 387: C. P. Bruter, Eléments de la Théorie des Matroïdes V, 138 pages. 1974. DM 18,–

Vol. 388: R. L. Lipsman, Group Representations. X, 166 pages 1974. DM 20,–

Vol. 389: M.-A. Knus et M. Ojanguren, Théorie de la Descente et Algèbres d' Azumaya. IV, 163 pages. 1974. DM 20,–

Vol. 390: P. A. Meyer, P. Priouret et F. Spitzer, Ecole d'Eté de Probabilités de Saint-Flour III – 1973. Edité par A. Badrikian et P.-L. Hennequin. VIII, 189 pages. 1974. DM 20,–

Vol. 391: J. Gray, Formal Category Theory: Adjointness for 2 Categories. XII, 282 pages. 1974. DM 24,–

Vol. 392: Géométrie Différentielle, Colloque, Santiago d Compostela, Espagne 1972. Edité par E. Vidal. VI, 225 pages 1974. DM 20,–

Vol. 393: G. Wassermann, Stability of Unfoldings. IX, 164 pages 1974. DM 20,–

Vol. 394: W. M. Patterson 3rd, Iterative Methods for the Solutio of a Linear Operator Equation in Hilbert Space – A Survey III, 183 pages. 1974. DM 20,–

Vol. 395: Numerische Behandlung nichtlinearer Integrodifferen tial- und Differentialgleichungen. Tagung 1973. Herausgegebe von R. Ansorge und W. Törnig. VII, 313 Seiten. 1974. DM 28,

Vol. 396: K. H. Hofmann, M. Mislove and A. Stralka, The Pontry agin Duality of Compact O-Dimensional Semilattices and it Applications. XVI, 122 pages. 1974. DM 18,–

Vol. 397: T. Yamada, The Schur Subgroup of the Brauer Group V, 159 pages. 1974. DM 18,–

Vol. 398: Théories de l'Information, Actes des Rencontres d Marseille-Luminy, 1973. Edité par J. Kampé de Fériet et C. Picard XII, 201 pages. 1974. DM 23,–